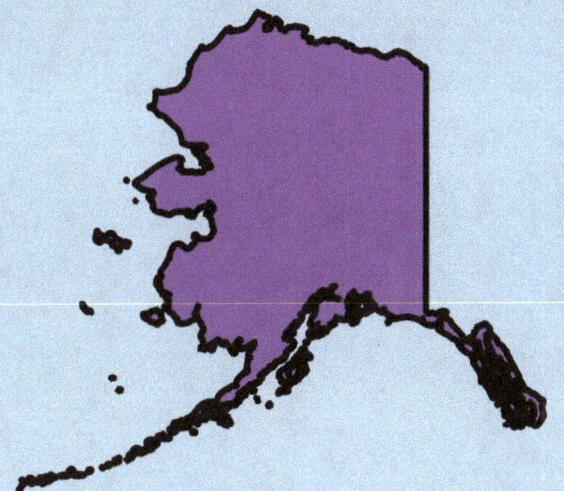
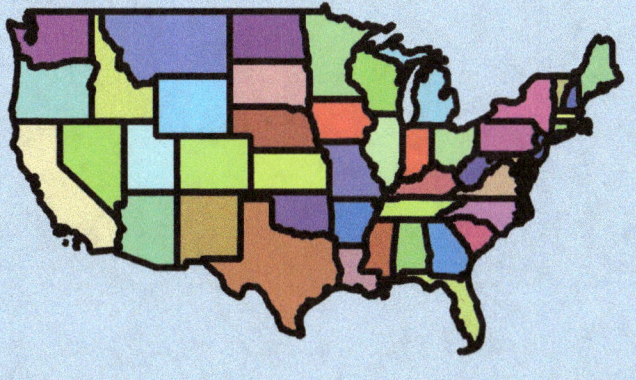

Cut and Glue Activity Book for Kids The United States

Copyrighted Materials

© M&M Baciu 2022
© OpenStreetMap Contributors
The data is available under the OpenStreetMap license
openstreetmap.org opendatacommons.org creativecommons.org

ALABAMA

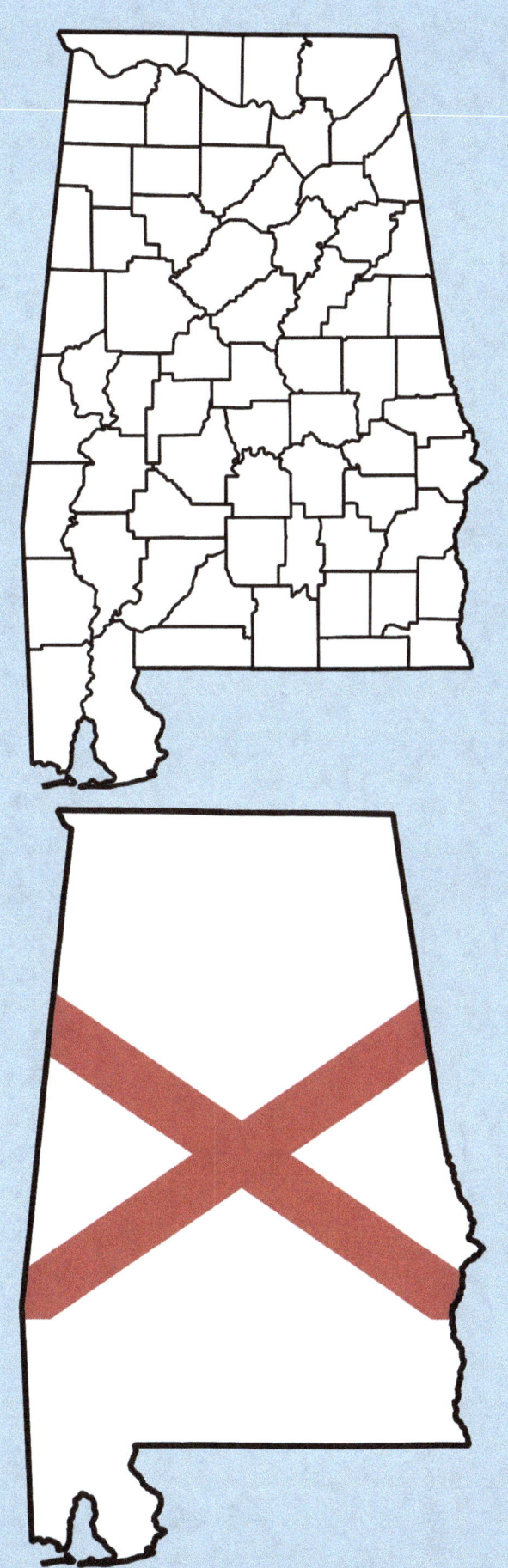

ALASKA

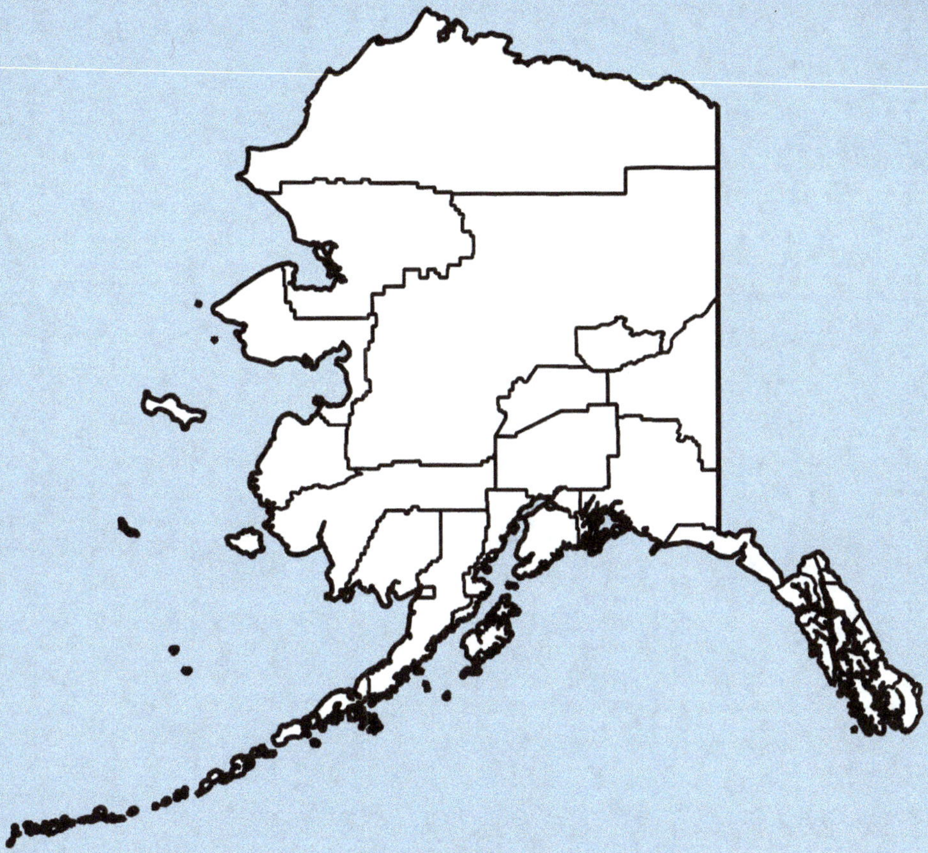

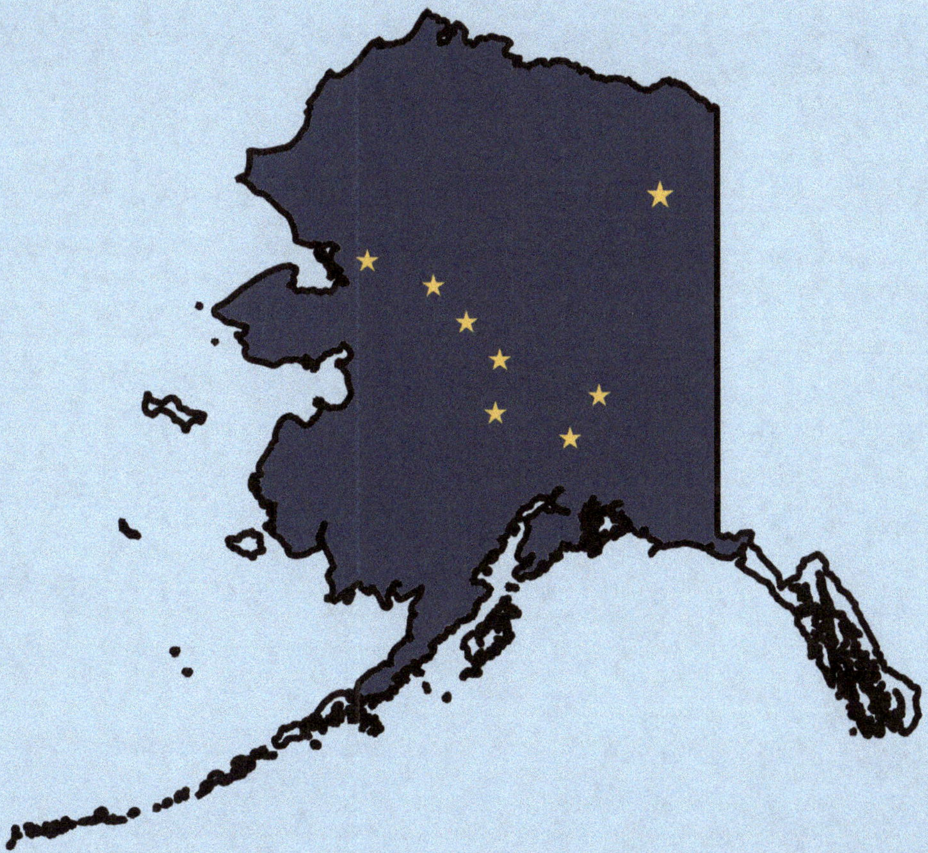

ARIZONA

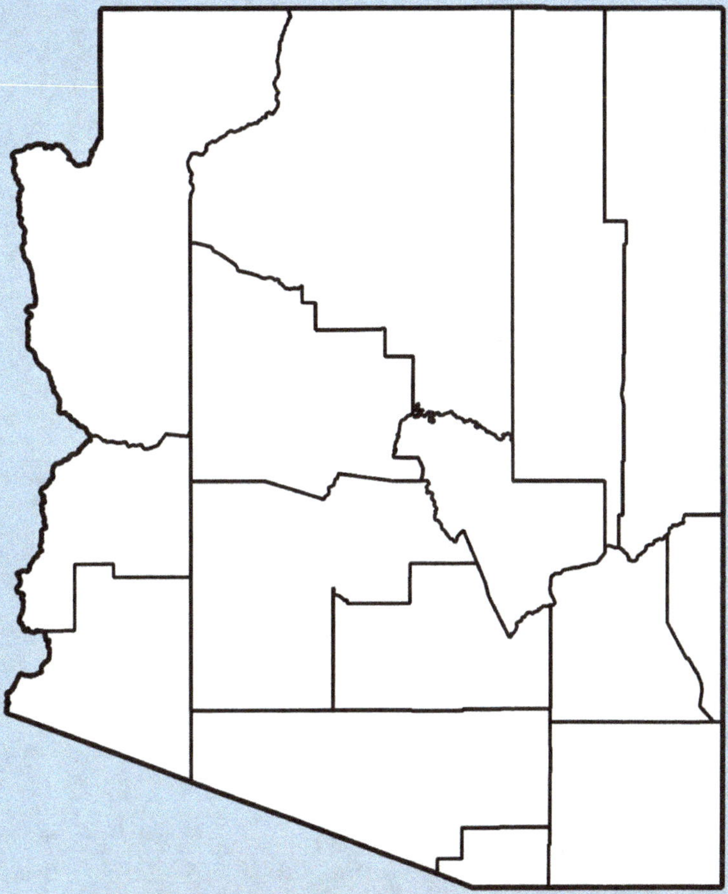

ARKANSAS

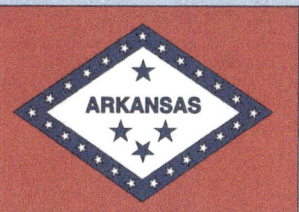

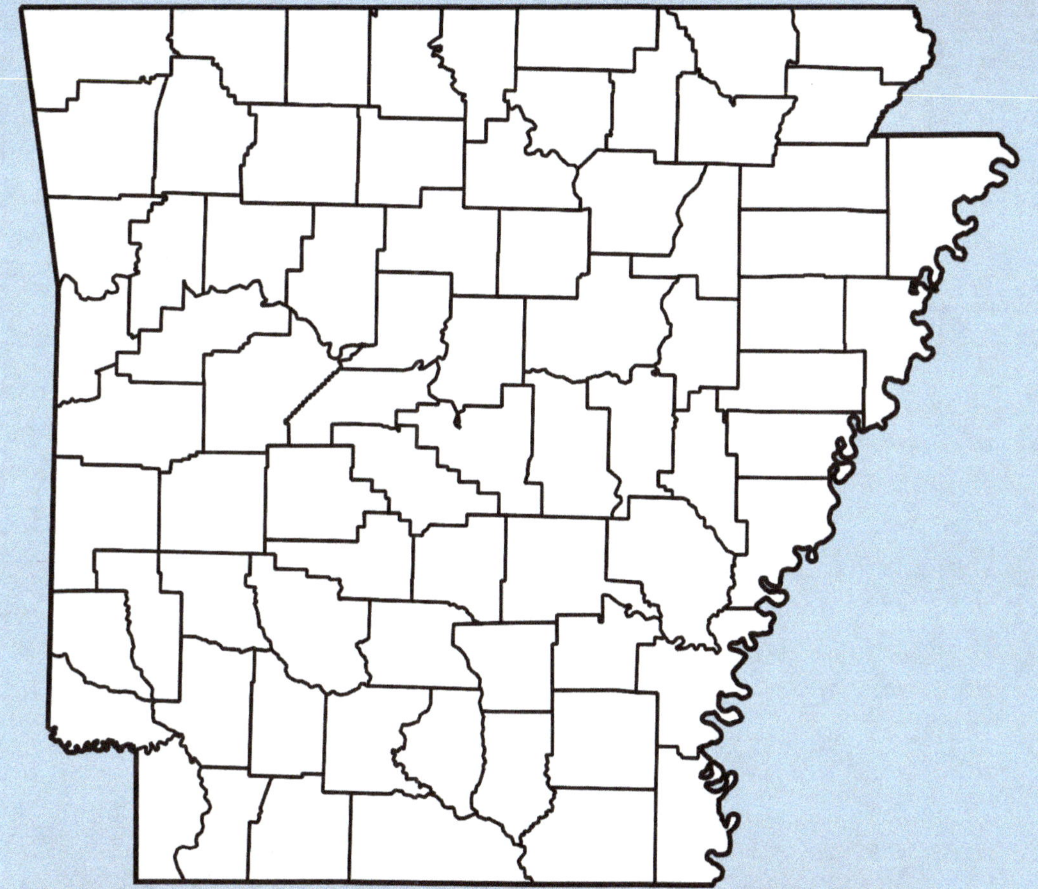

CALIFORNIA

COLORADO

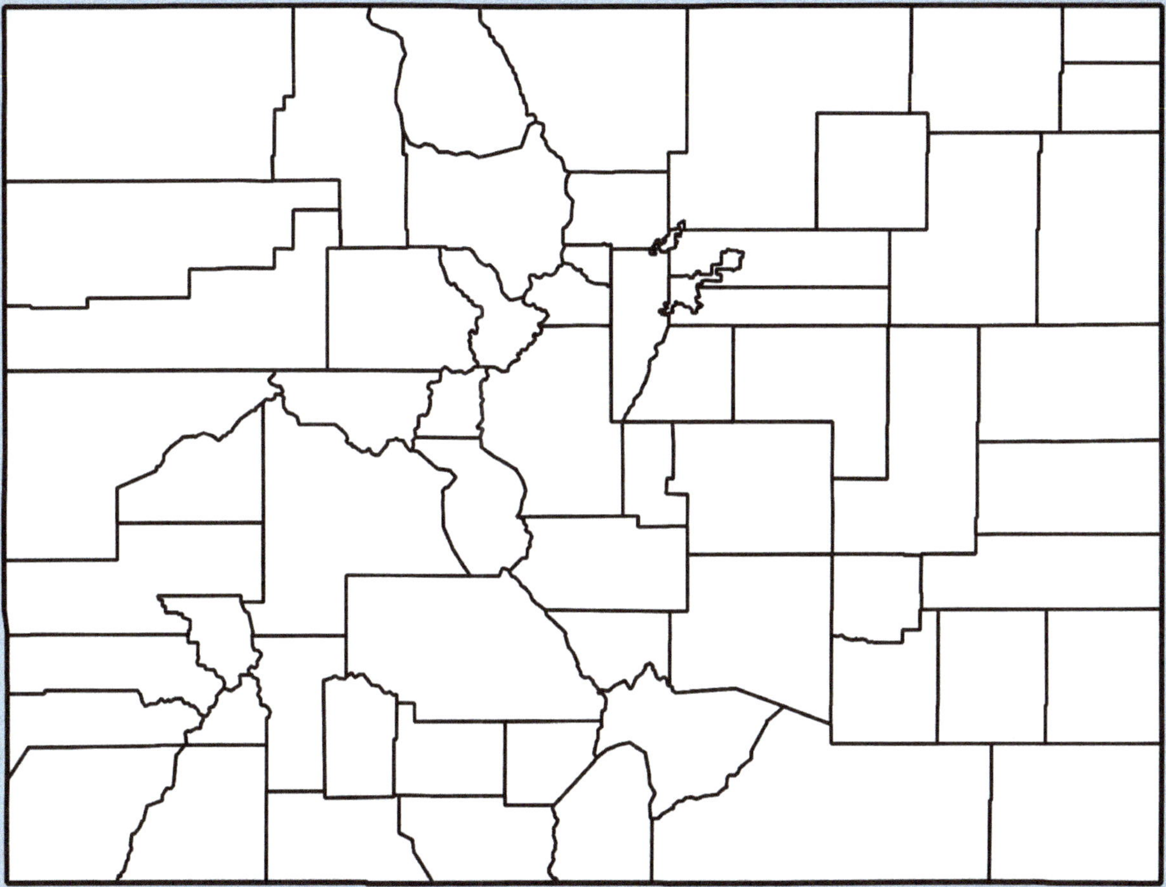

CONNECTICUT

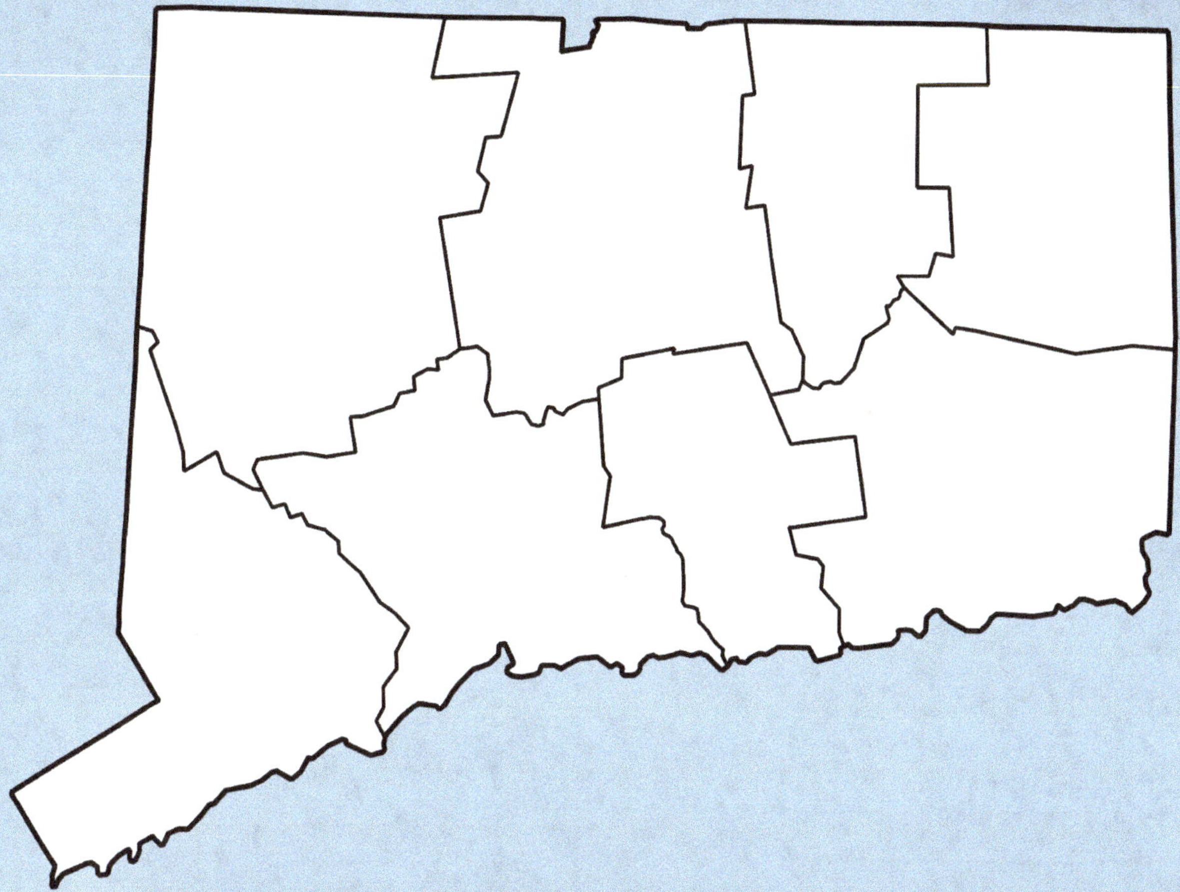

DELAWARE

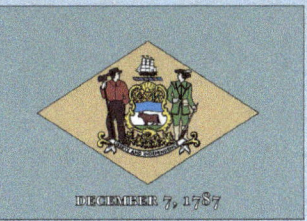

FLORIDA

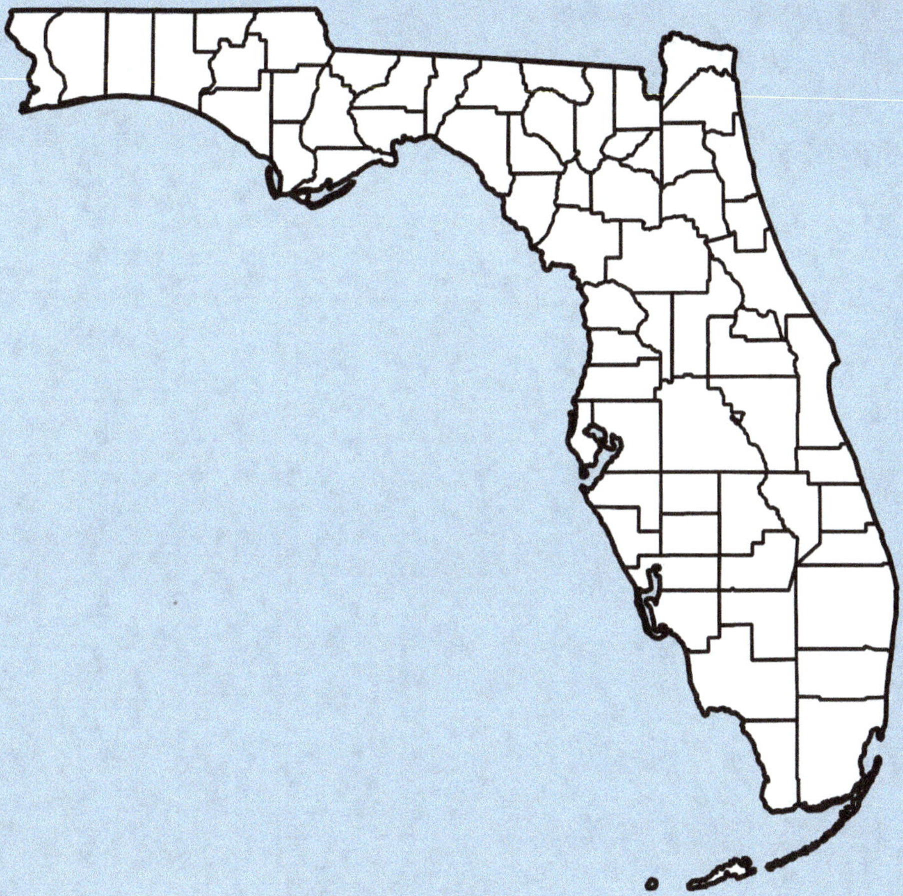

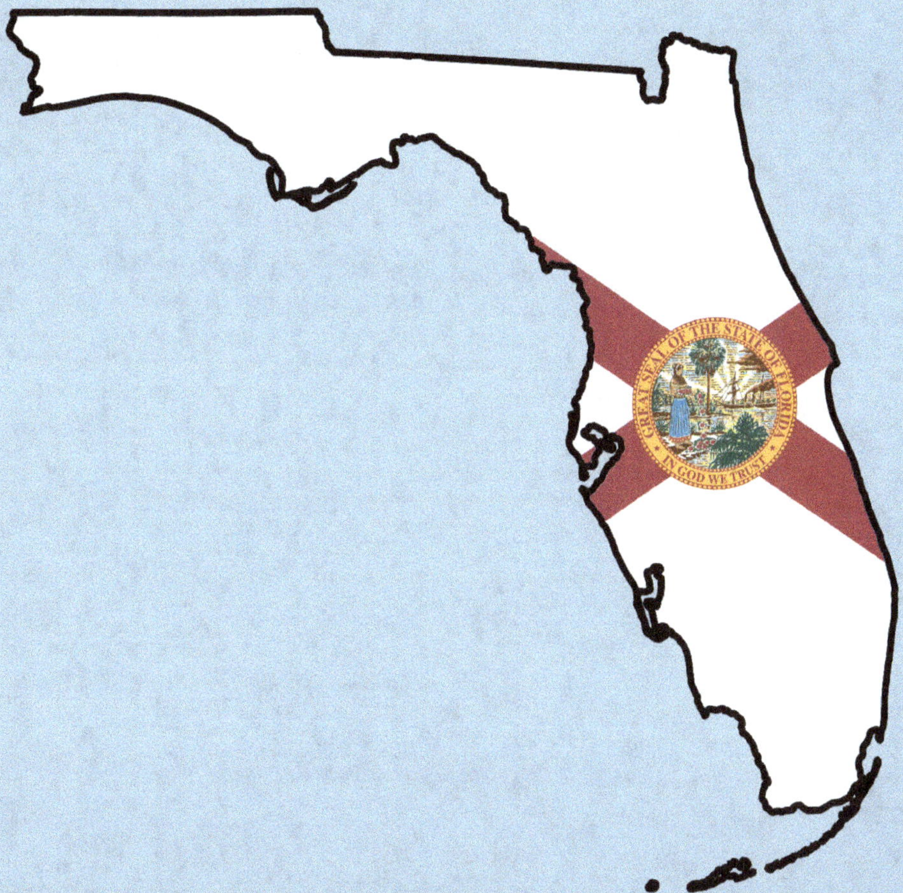

GEORGIA

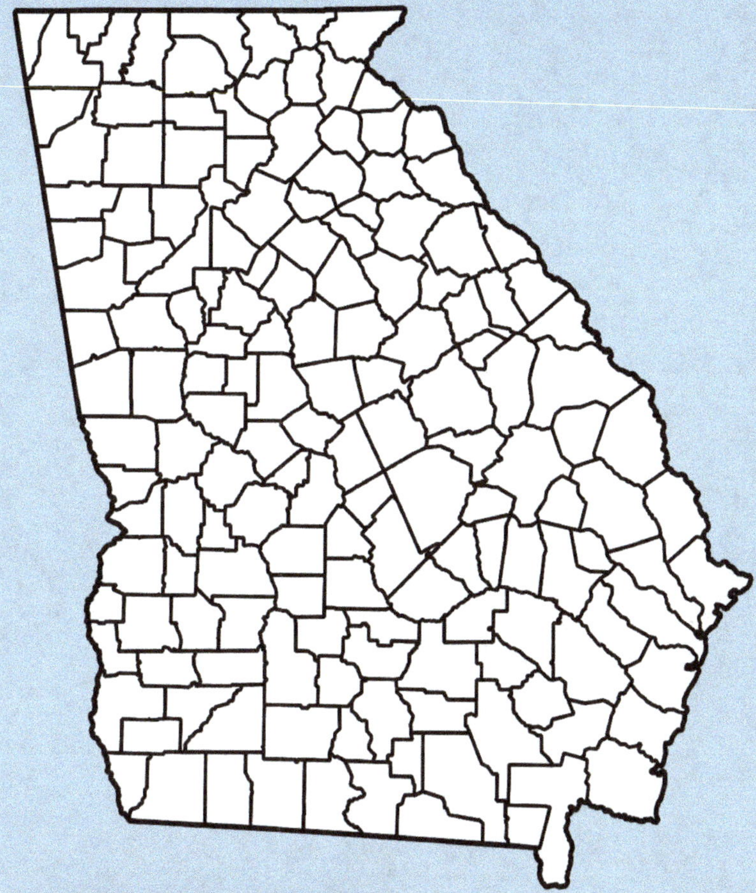

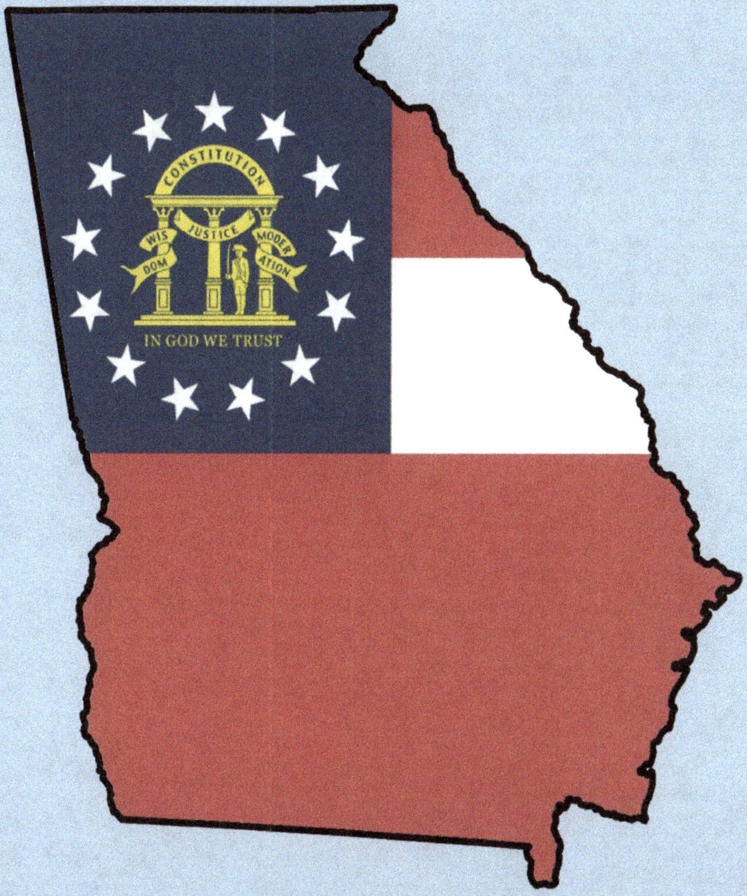

HAWAII

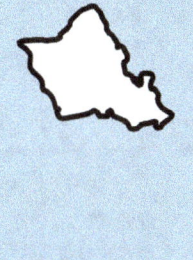

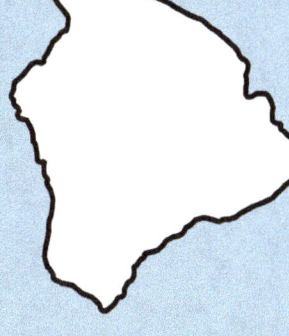

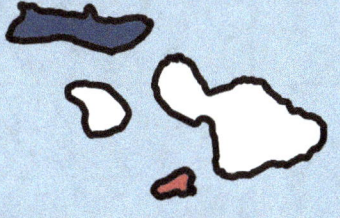

IDAHO

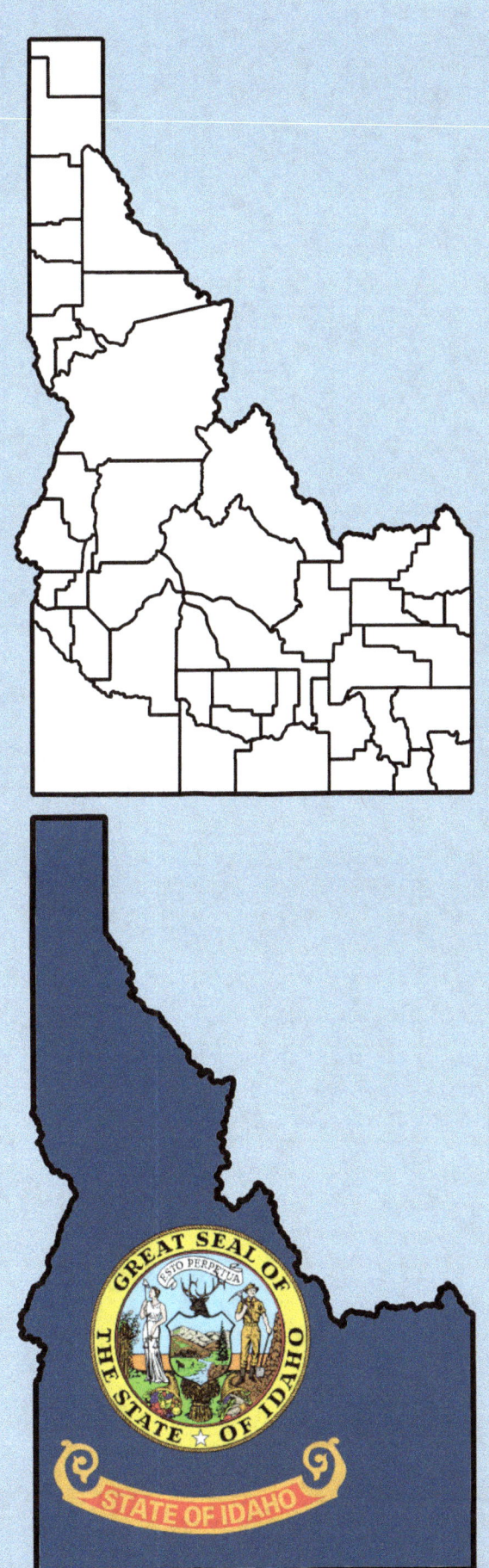

ILLINOIS

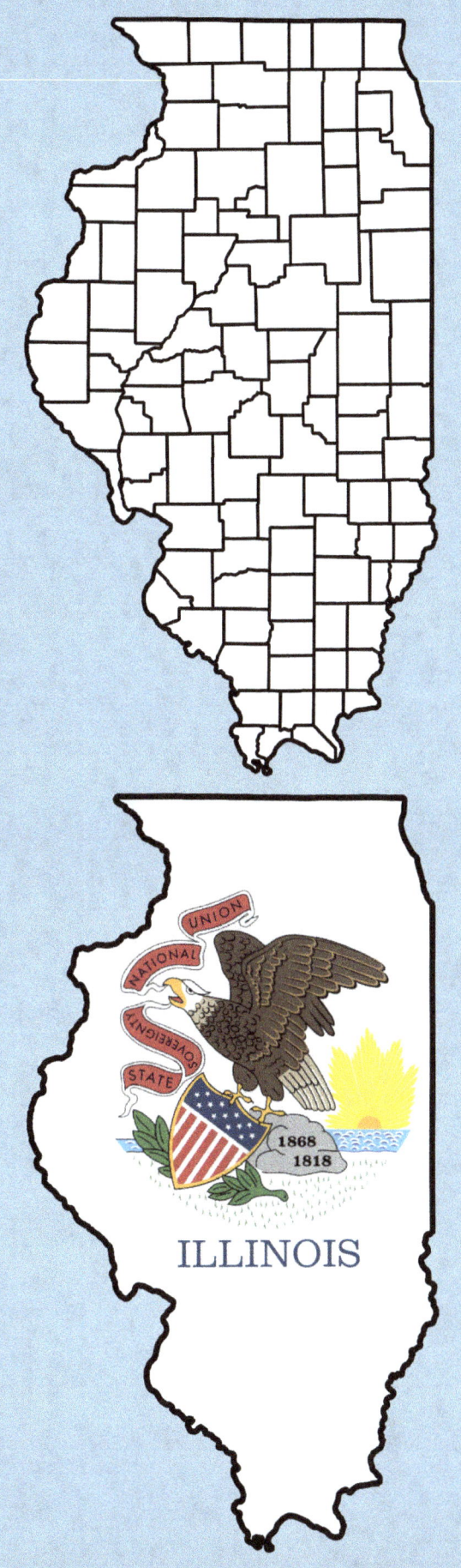

INDIANA

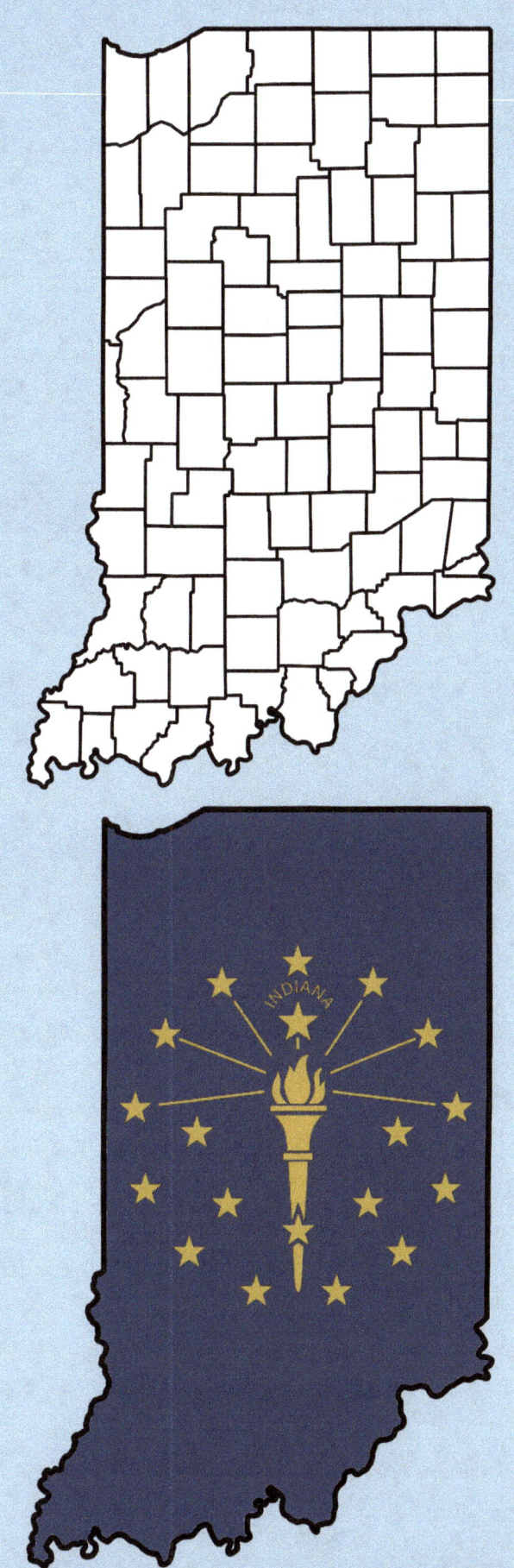

IOWA

IOWA

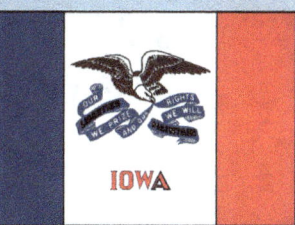

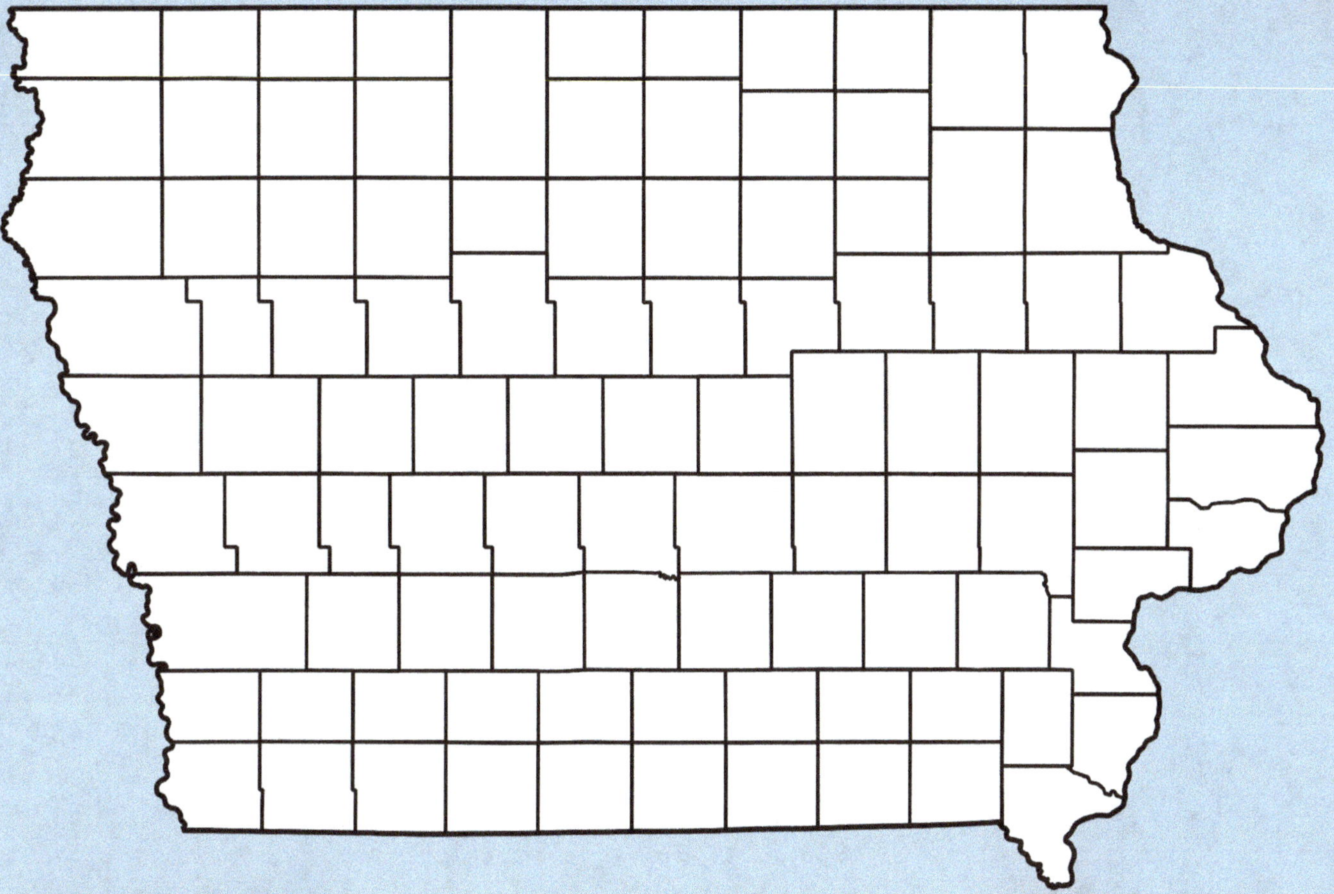

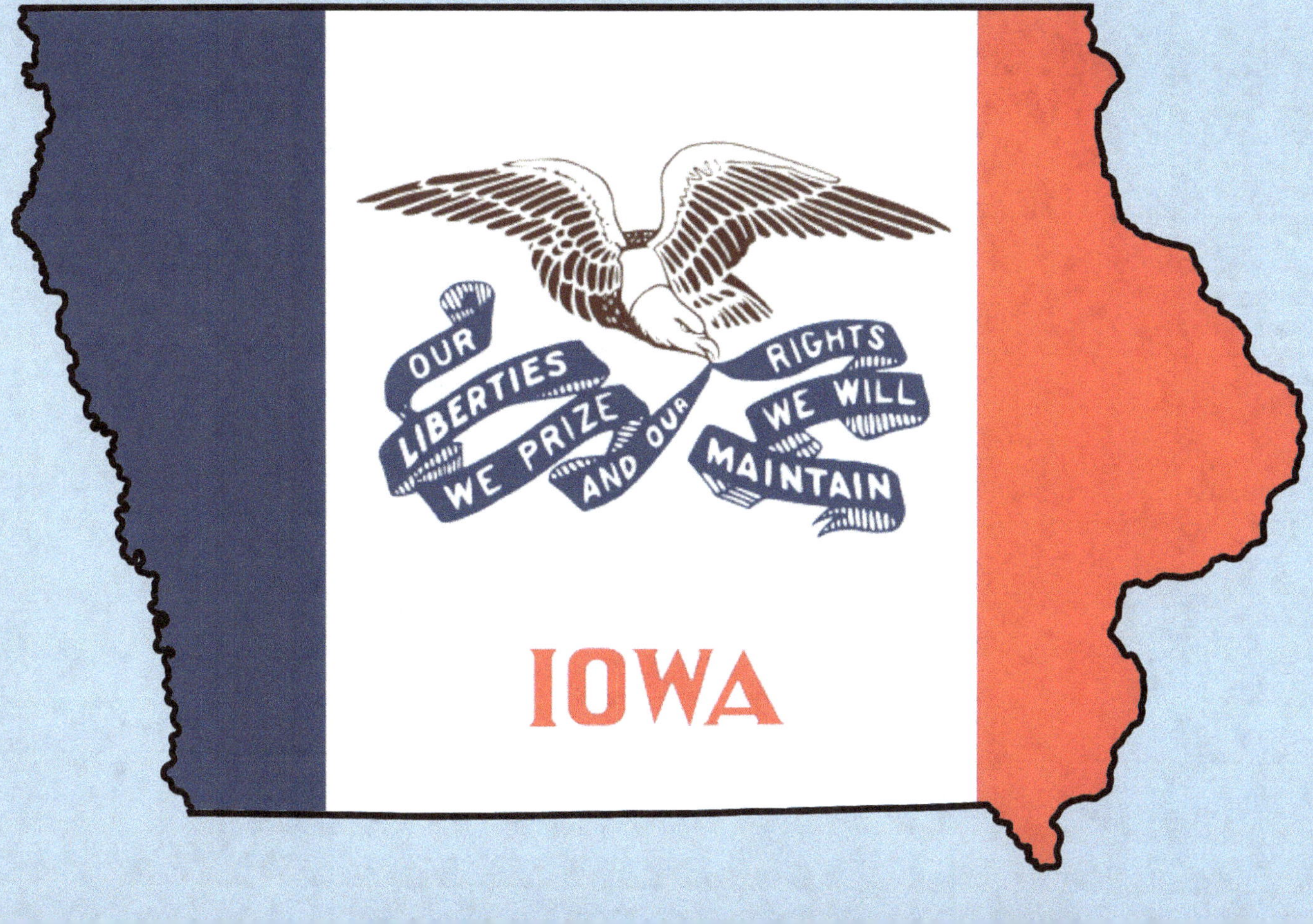
OUR LIBERTIES WE PRIZE
AND OUR RIGHTS WE WILL MAINTAIN
IOWA

KANSAS

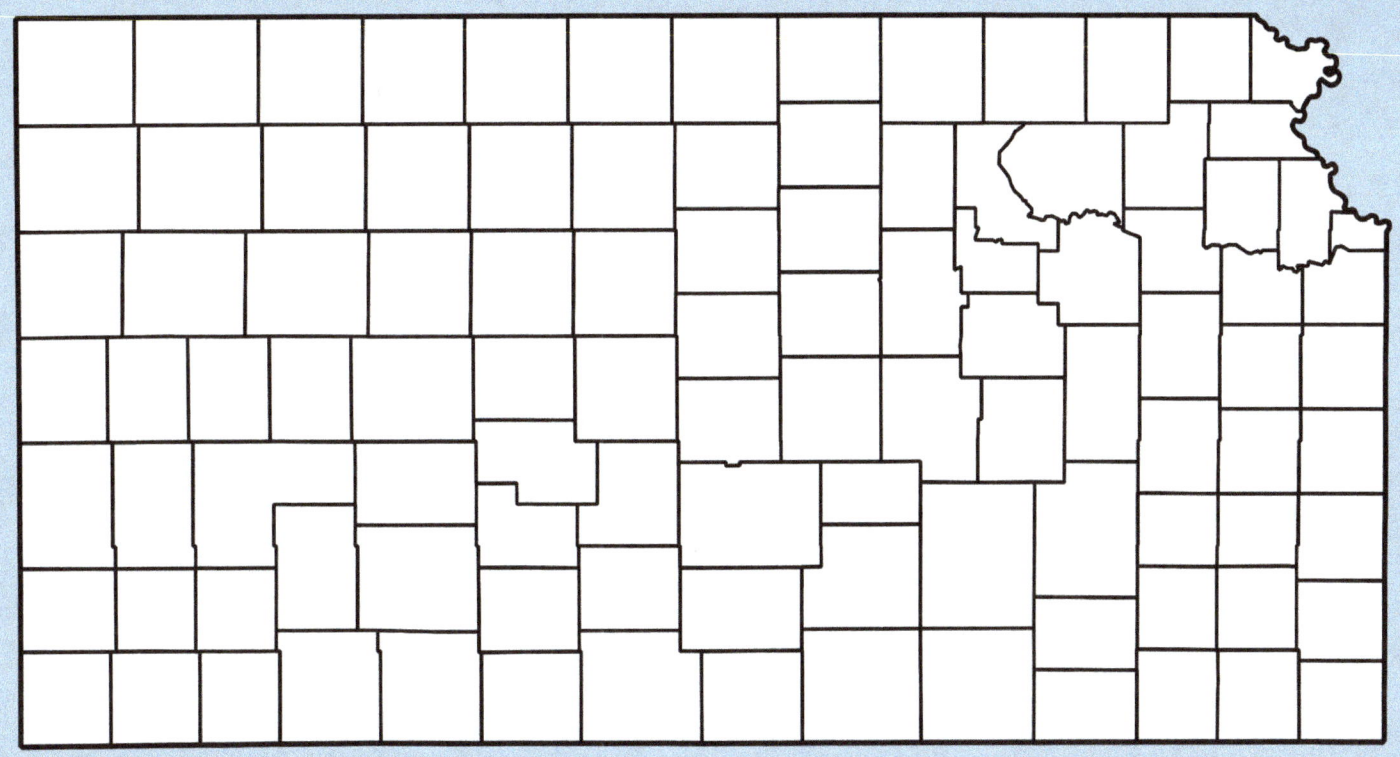

KENTUCKY

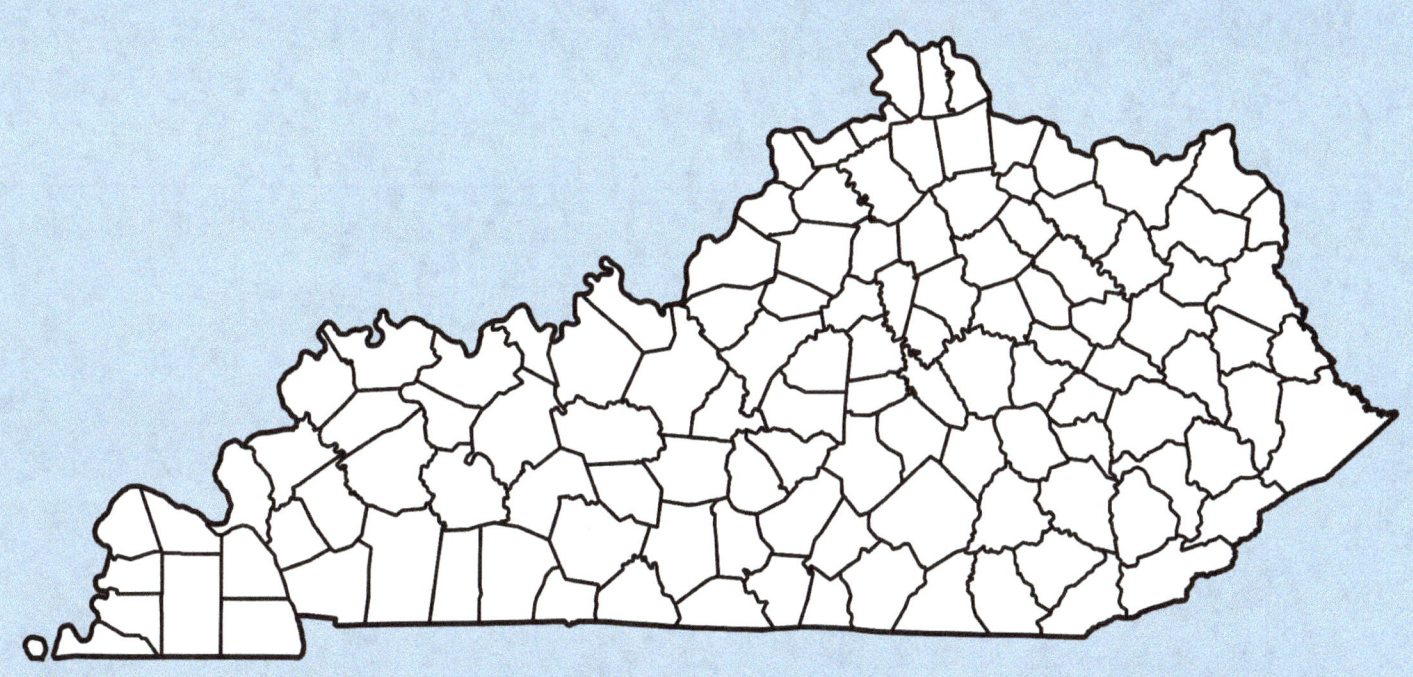
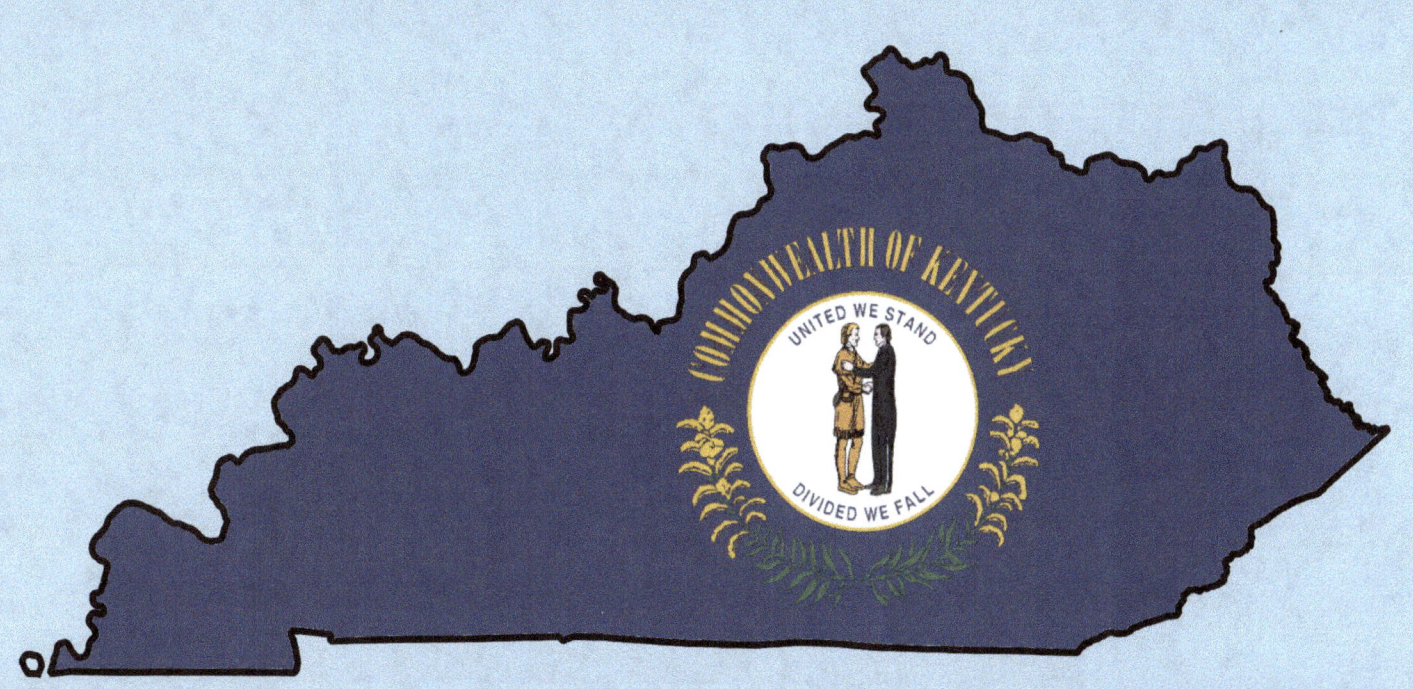

LOUISIANA

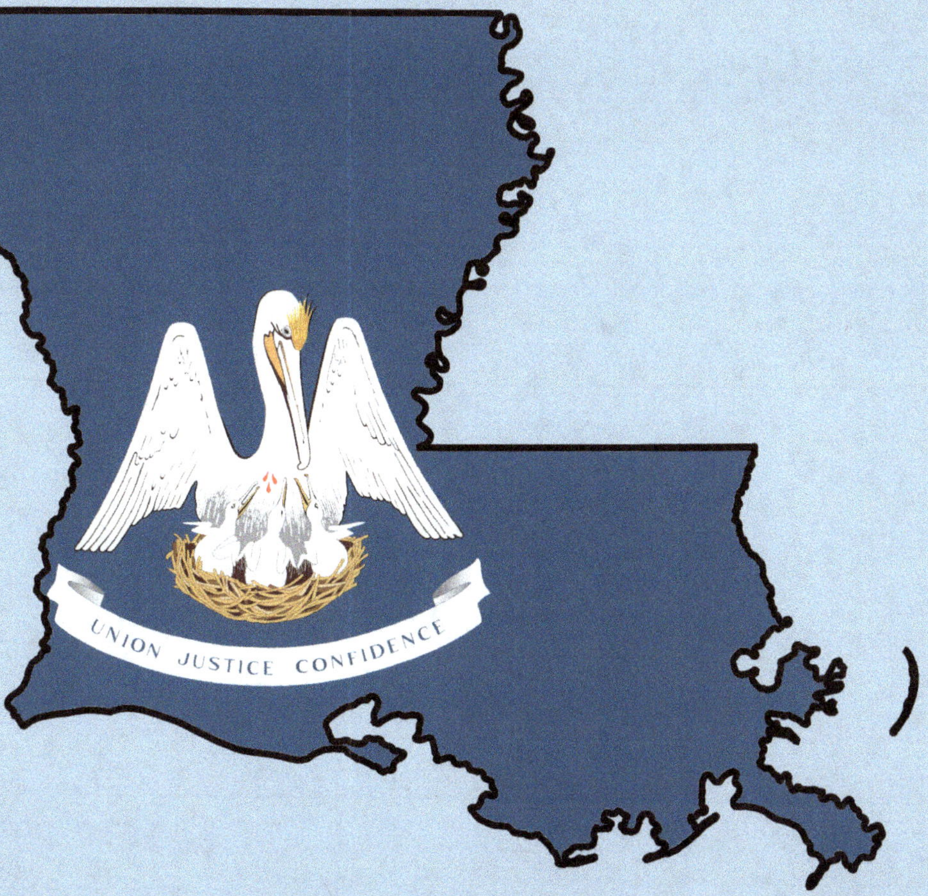

MAINE

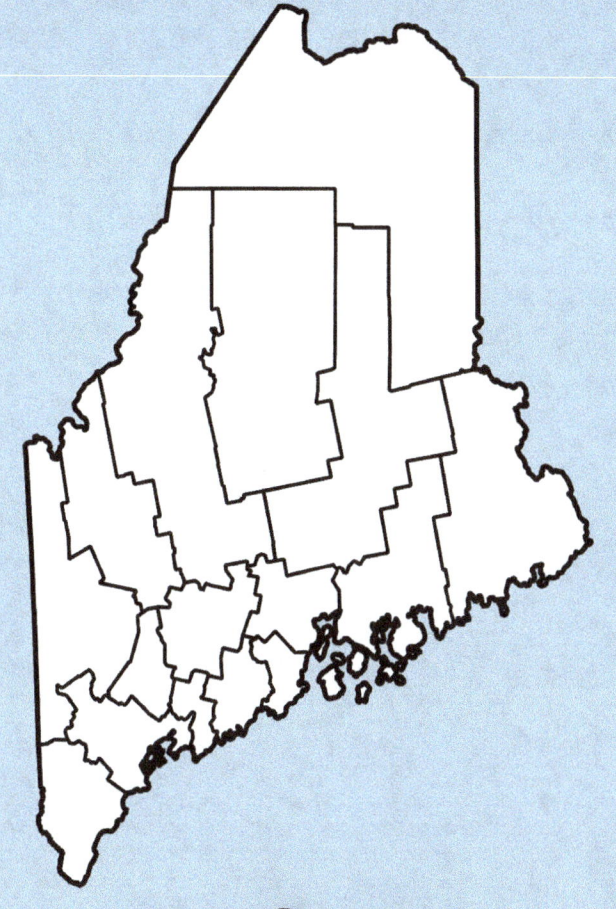

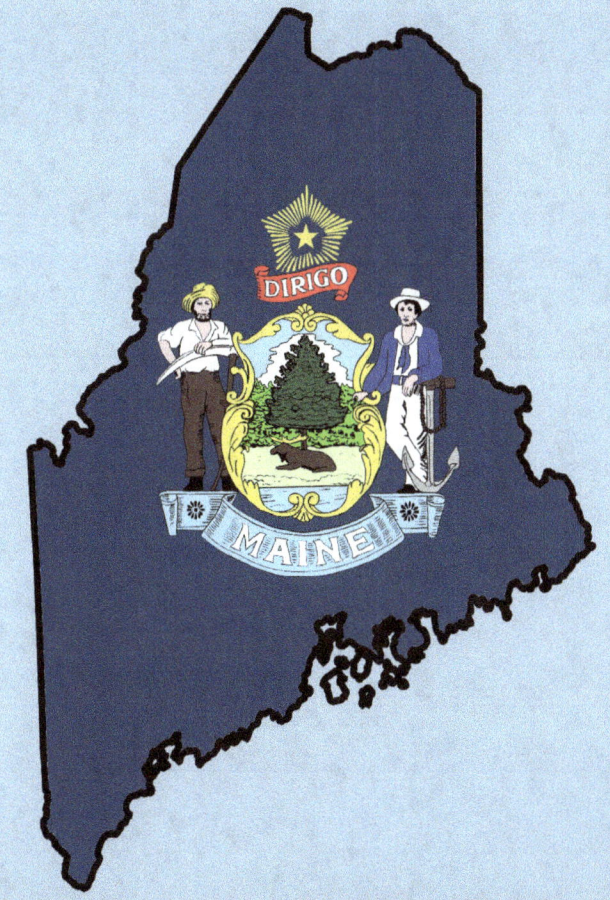

MARYLAND

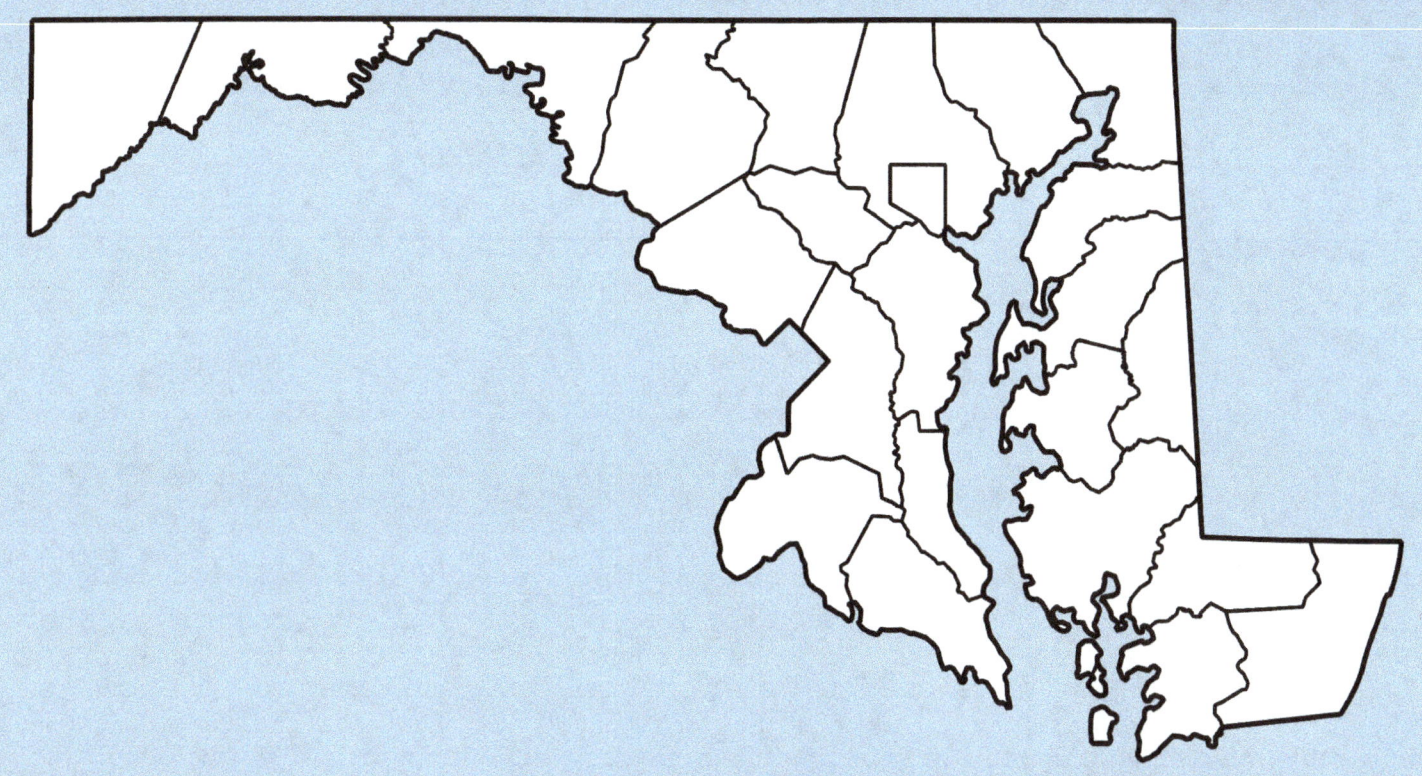

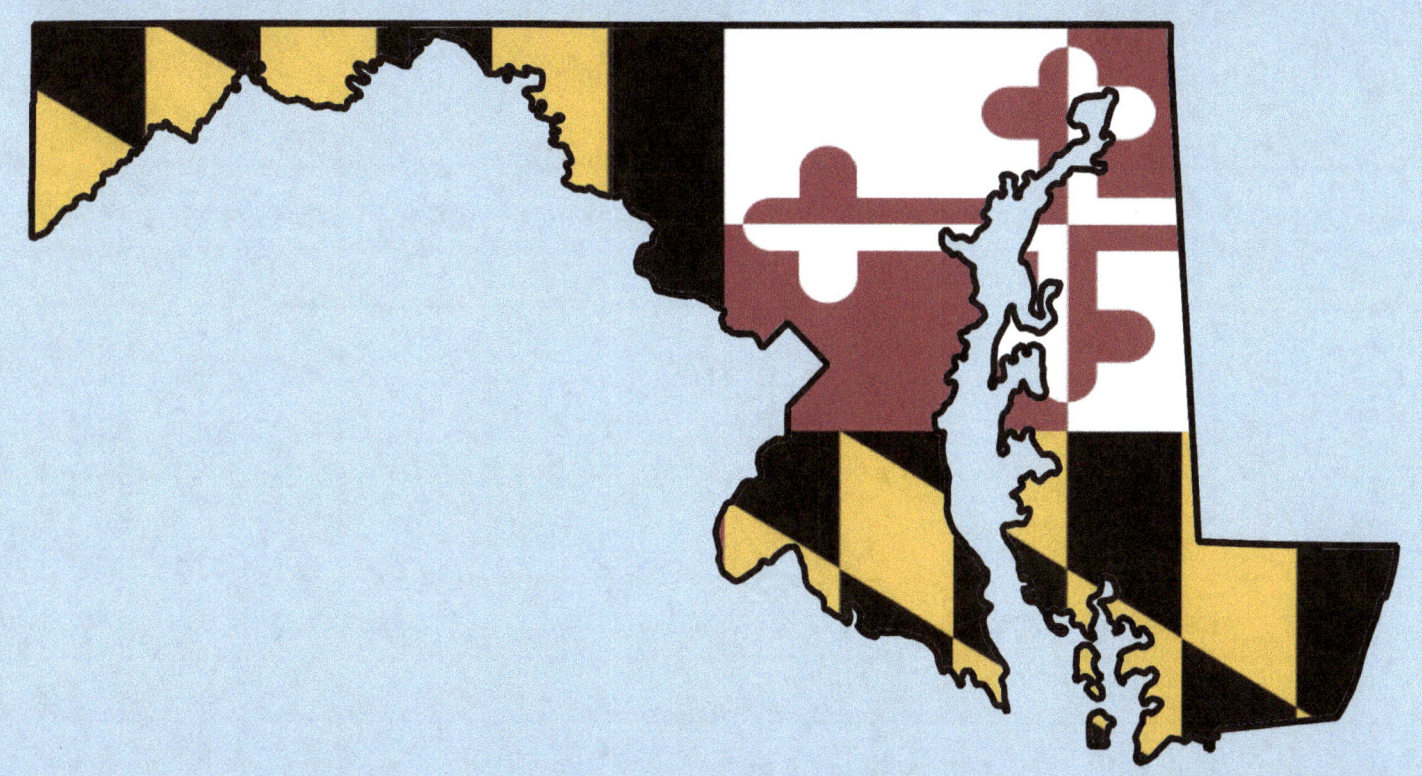

MASSACHUSETTS

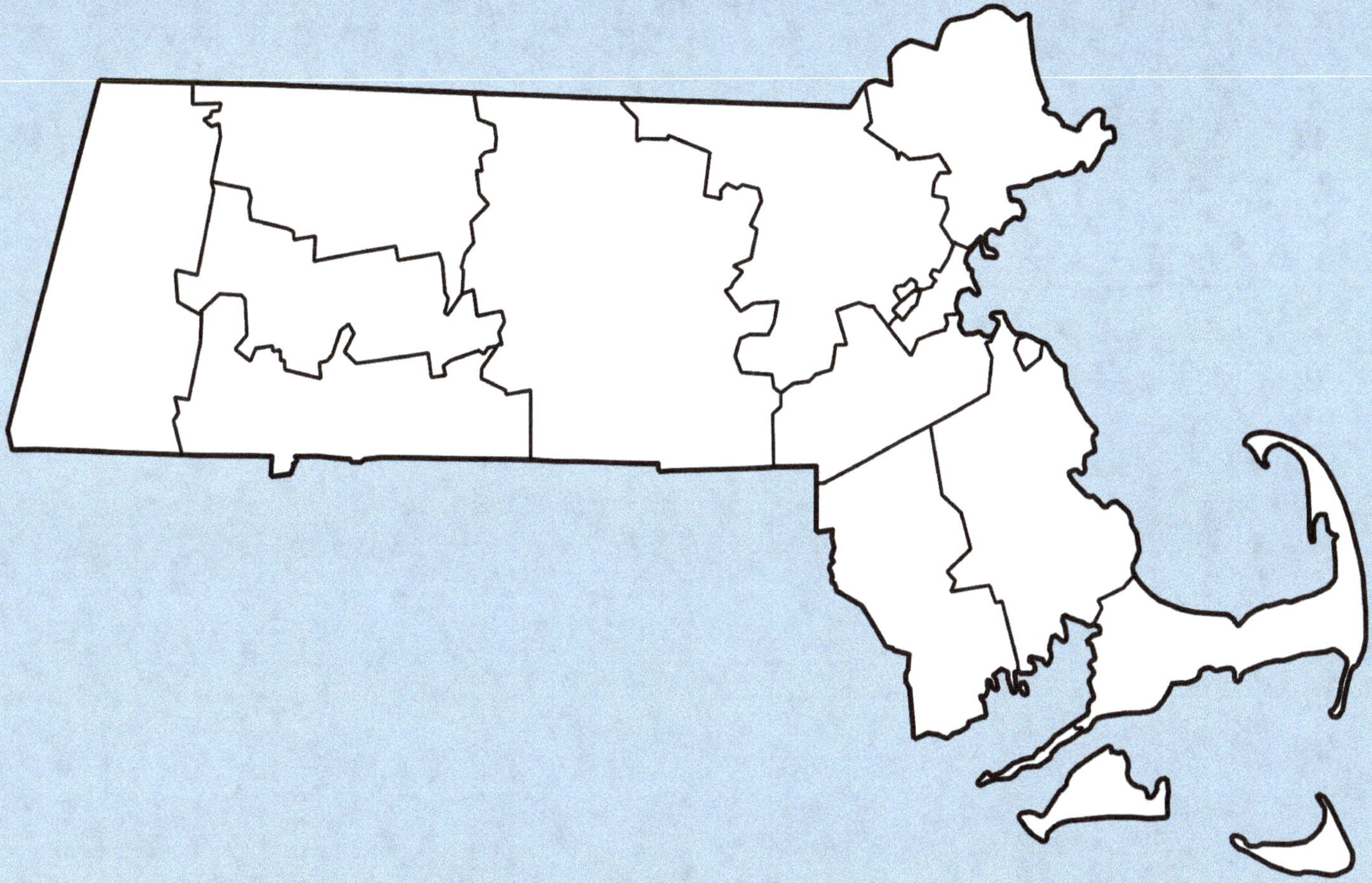

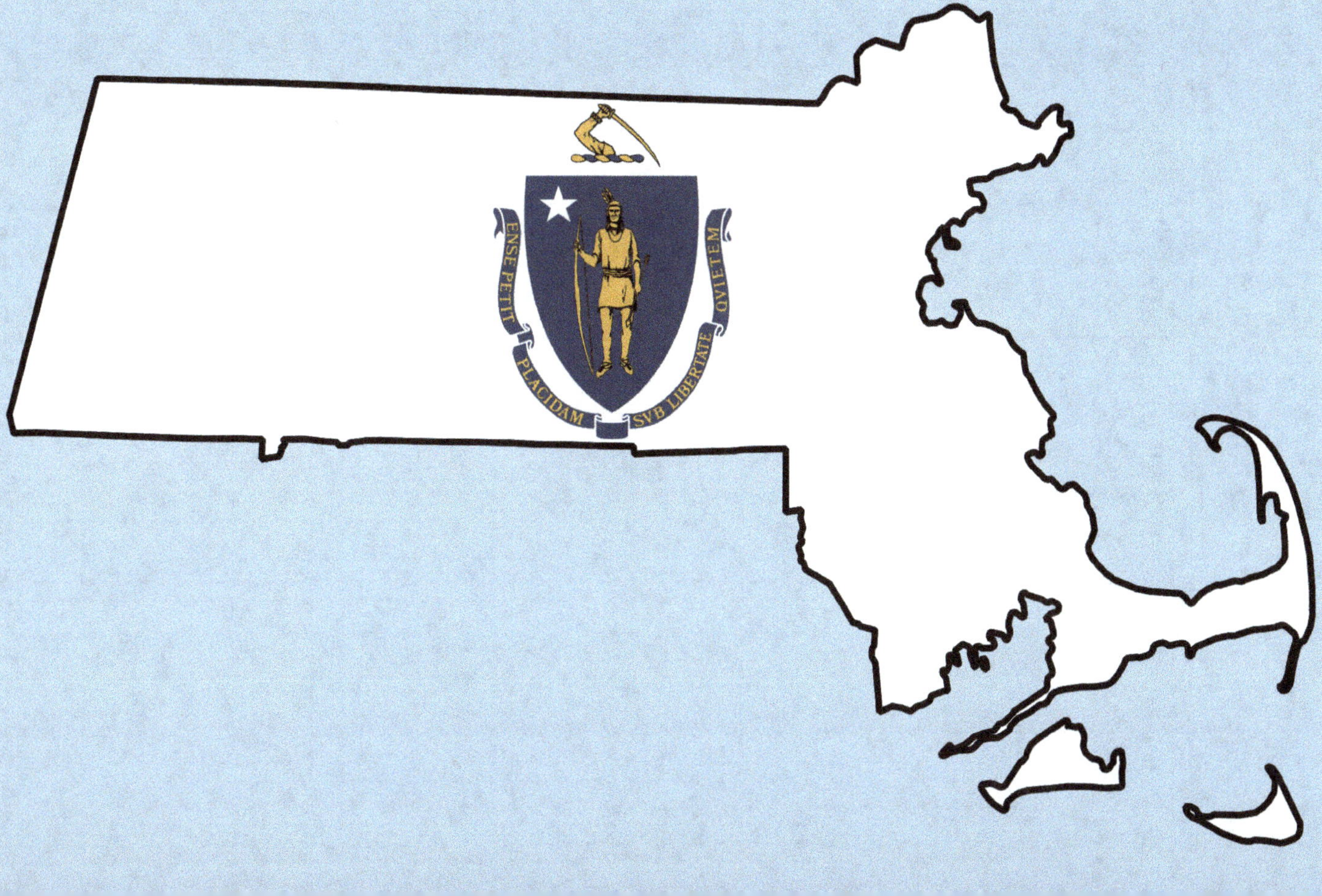

MICHIGAN

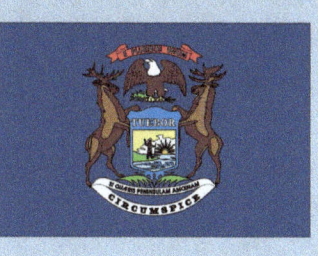

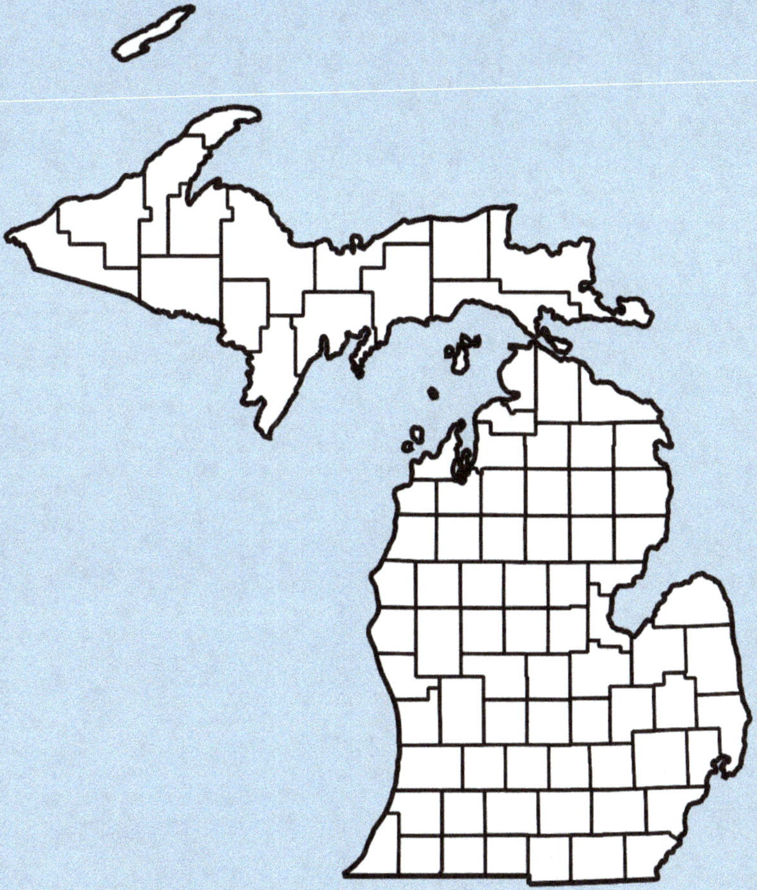

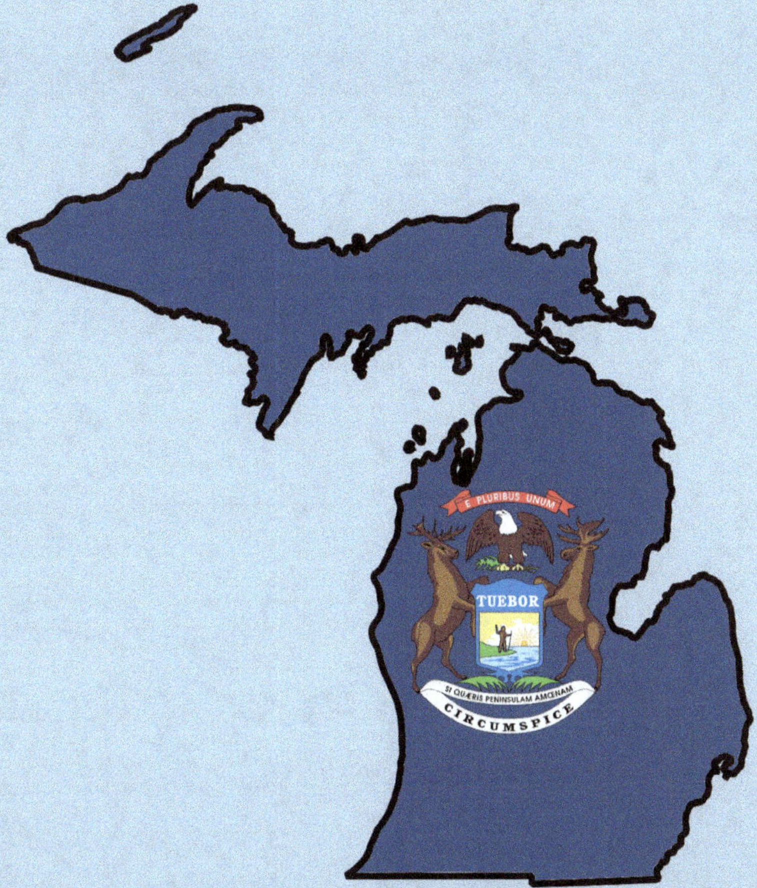
E PLURIBUS UNUM
TUEBOR
SI QUÆRIS PENINSULAM AMŒNAM
CIRCUMSPICE

MINNESOTA

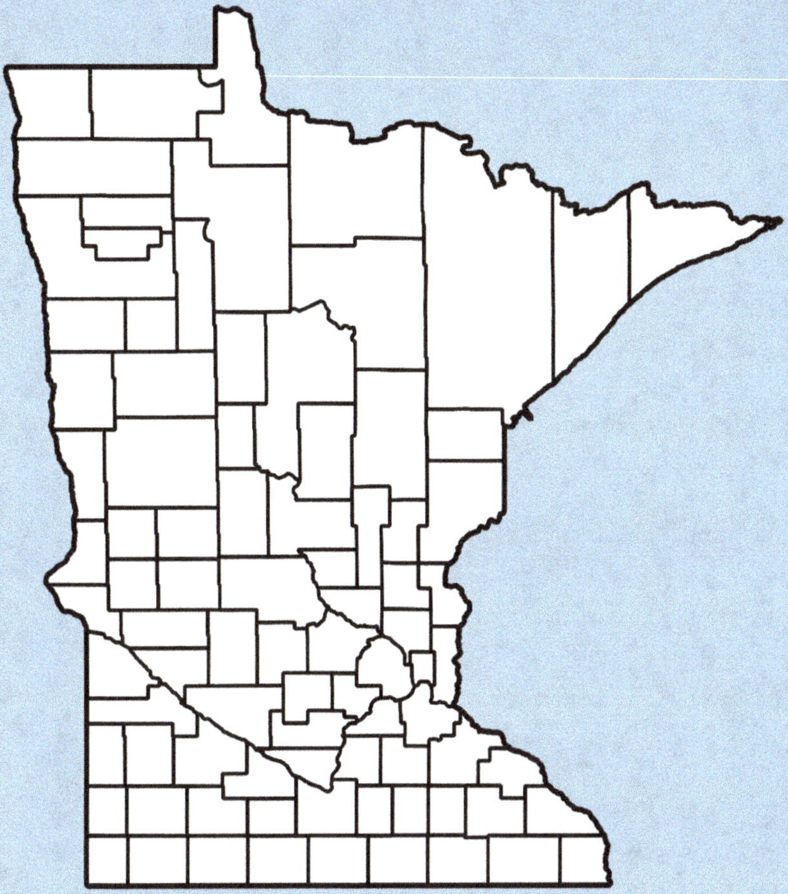

MISSISSIPPI

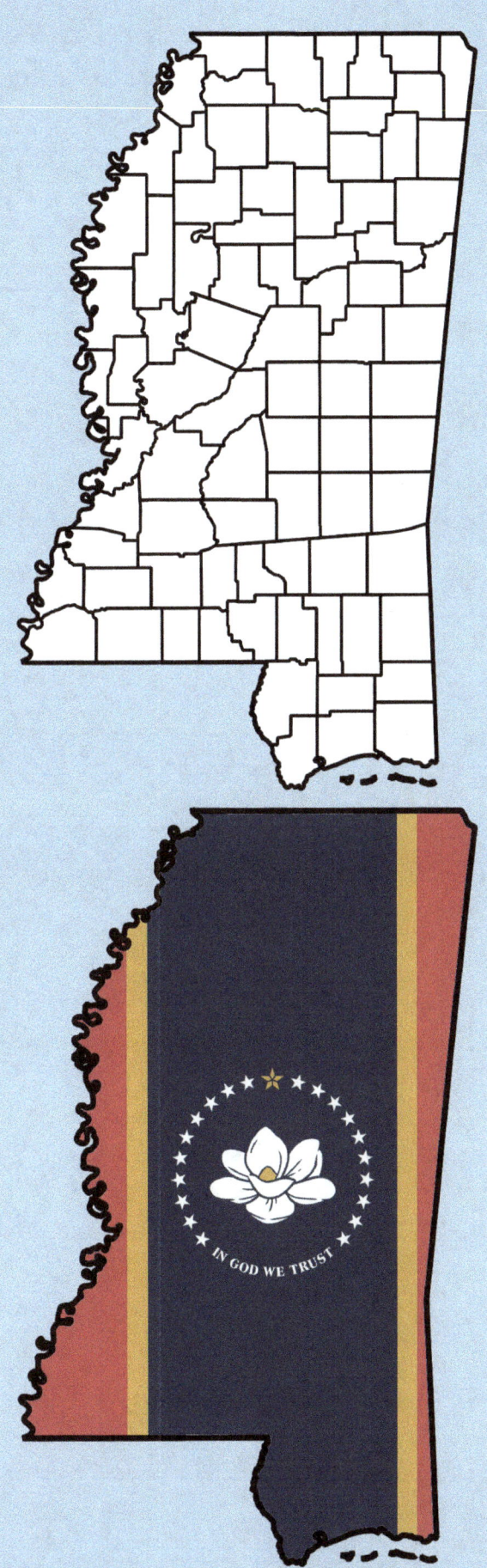

MISSOURI

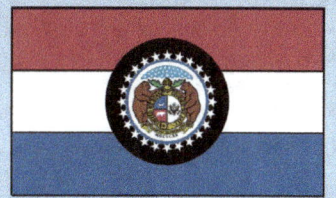

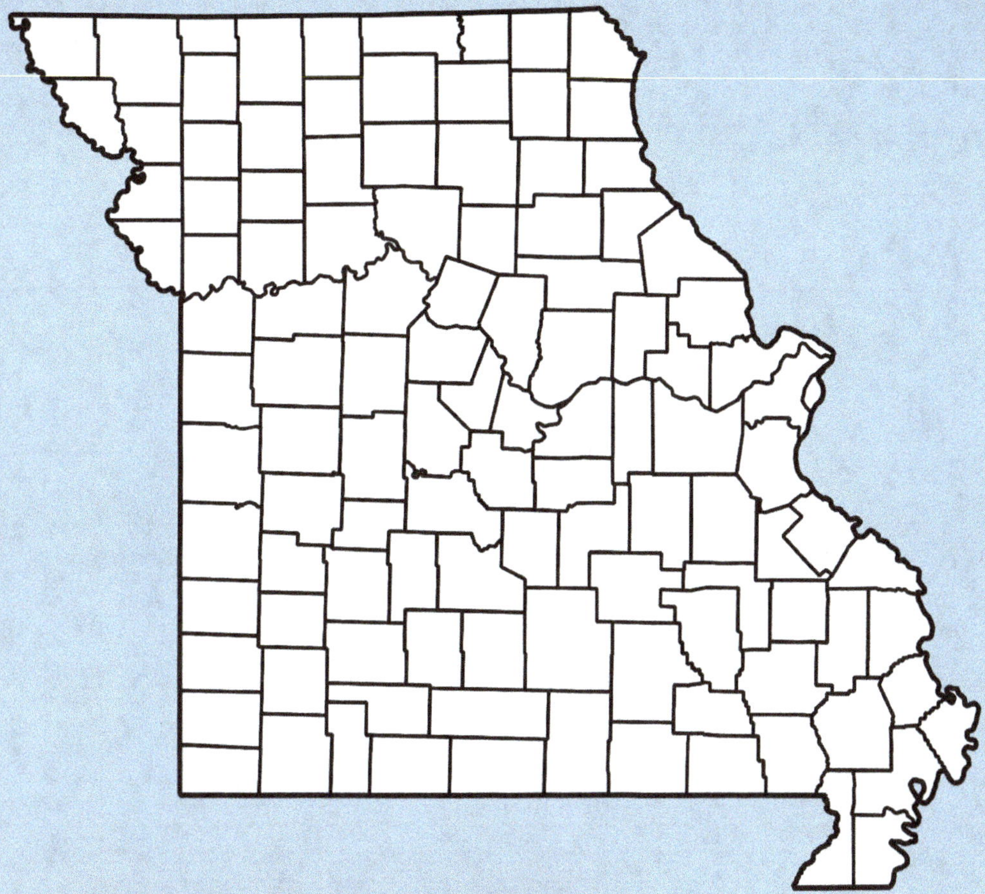

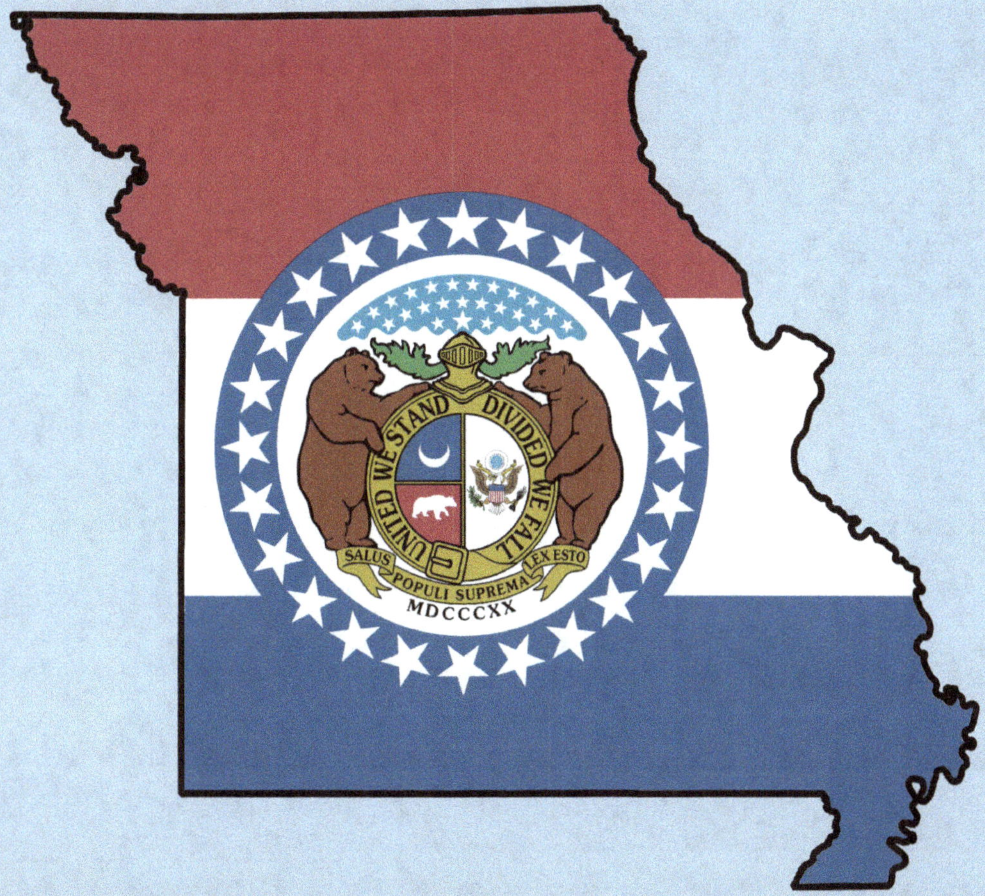
UNITED WE STAND DIVIDED WE FALL
SALUS POPULI SUPREMA LEX ESTO
MDCCCXX

MONTANA

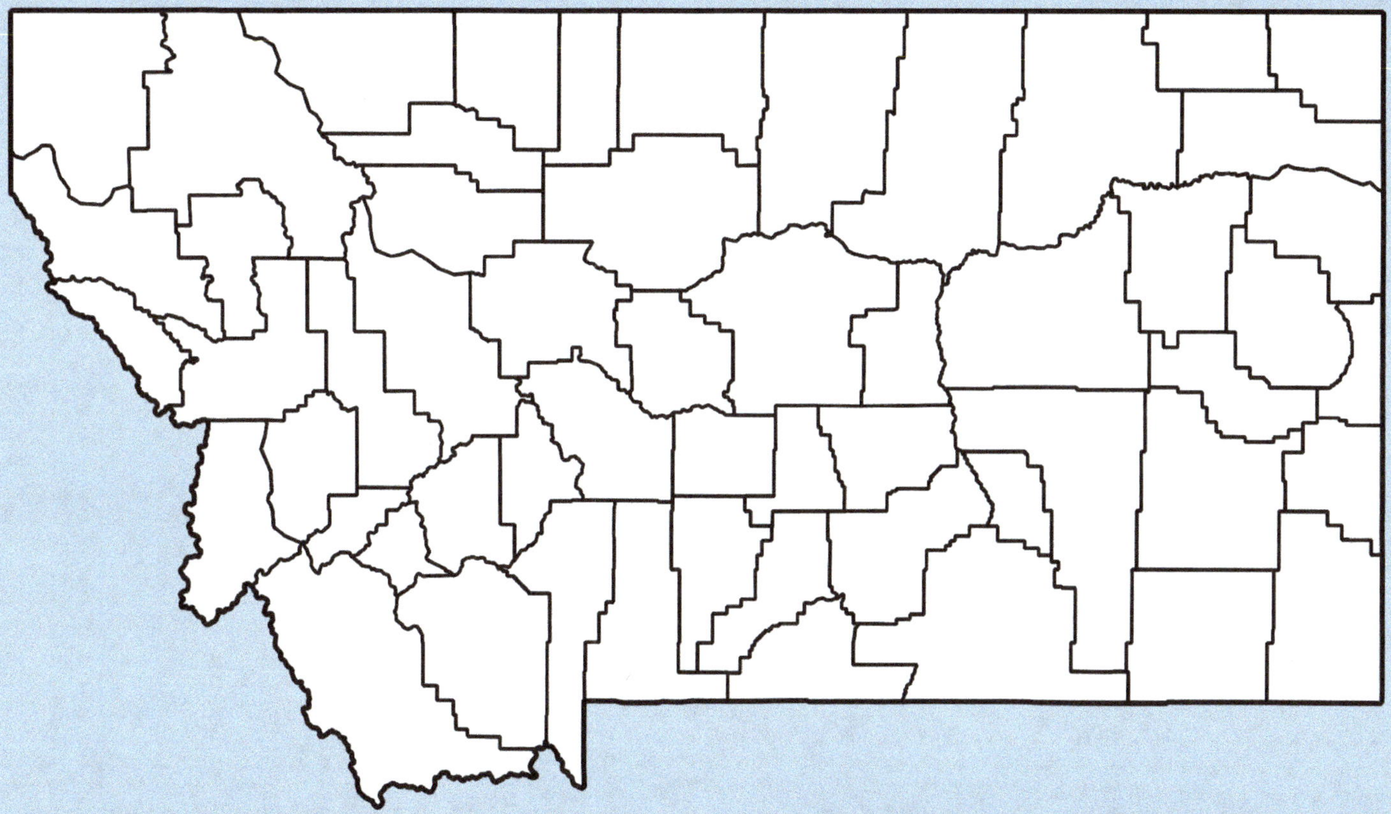

NEBRASKA

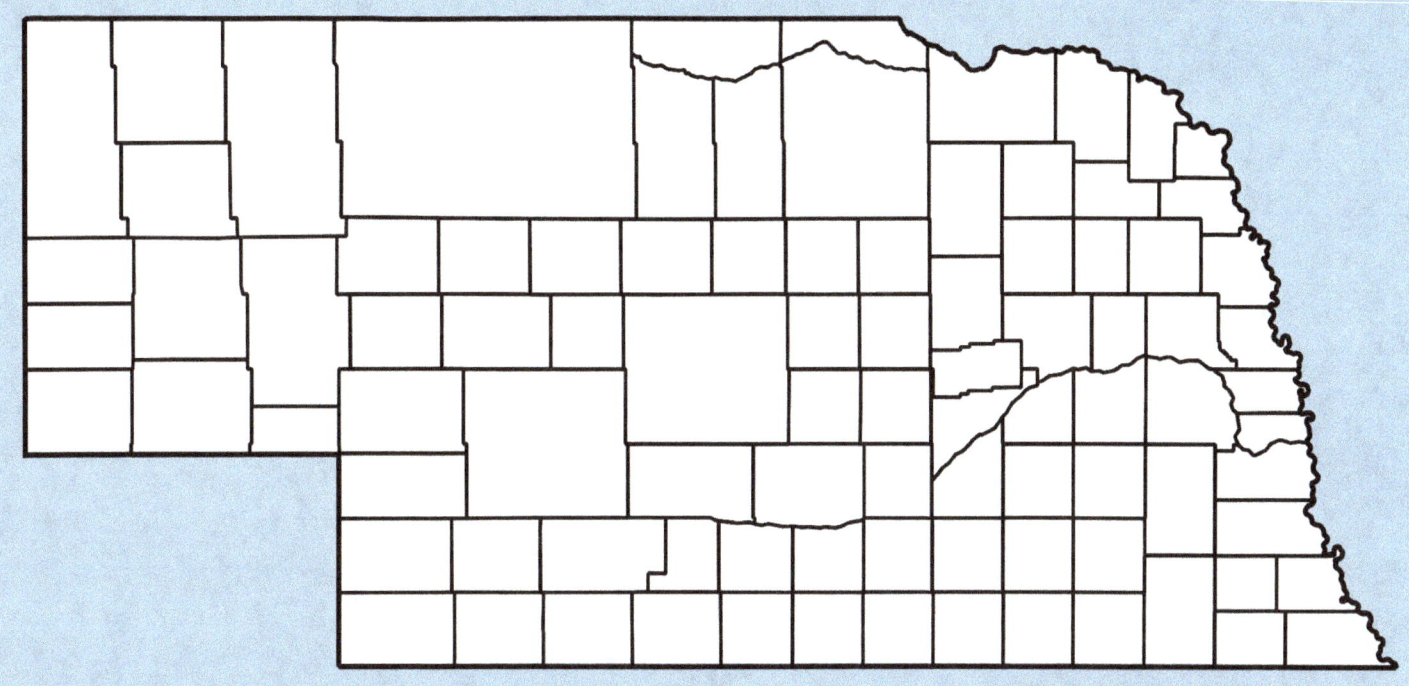

NEVADA

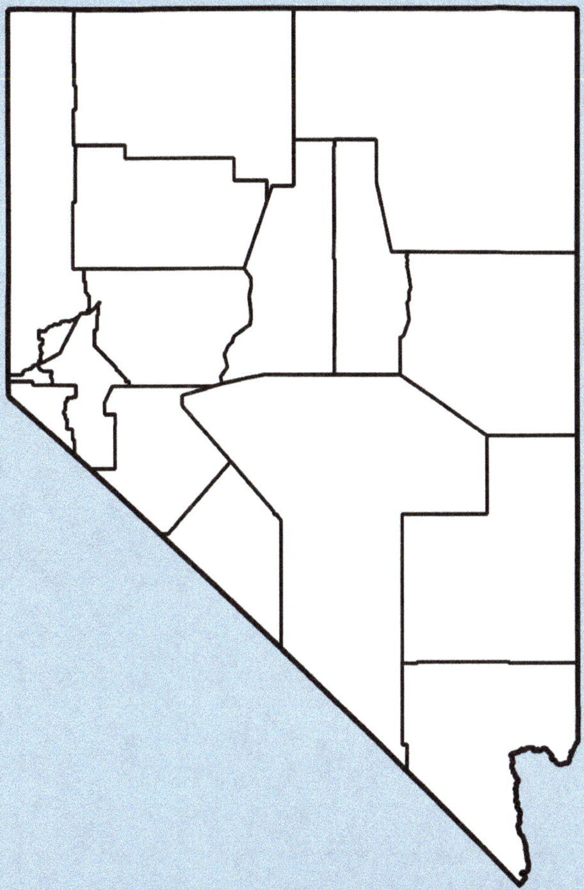

NEW HAMPSHIRE

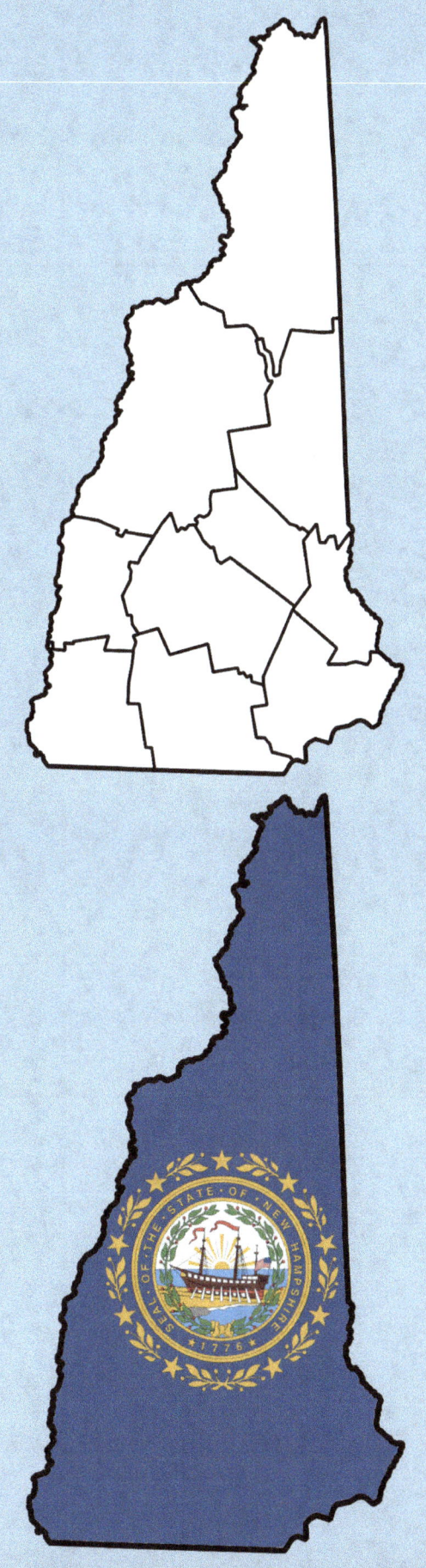

NEW JERSEY

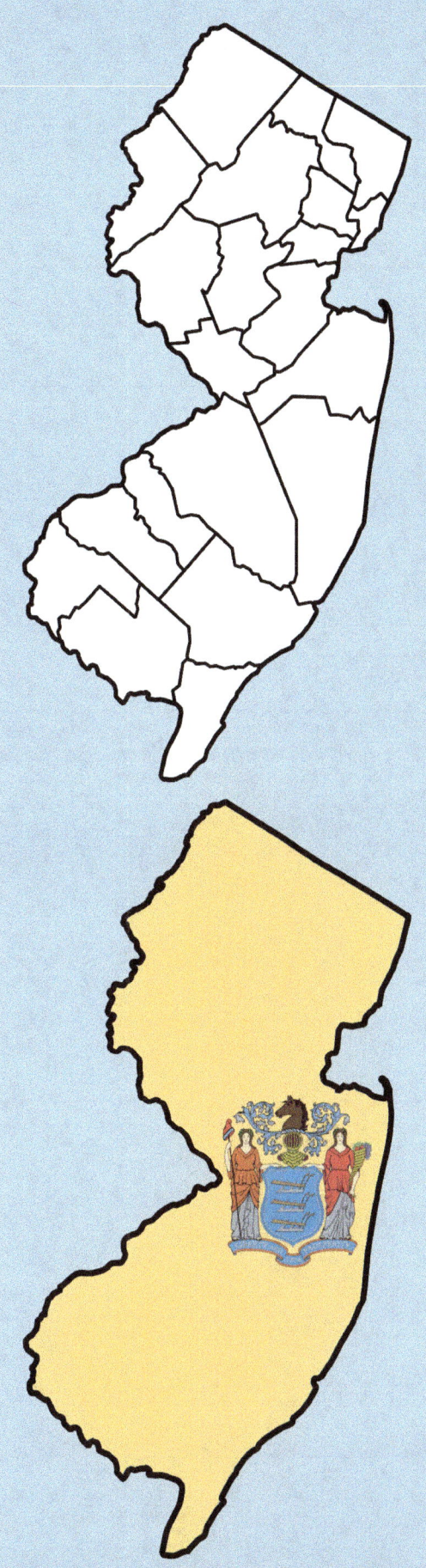

NEW MEXICO

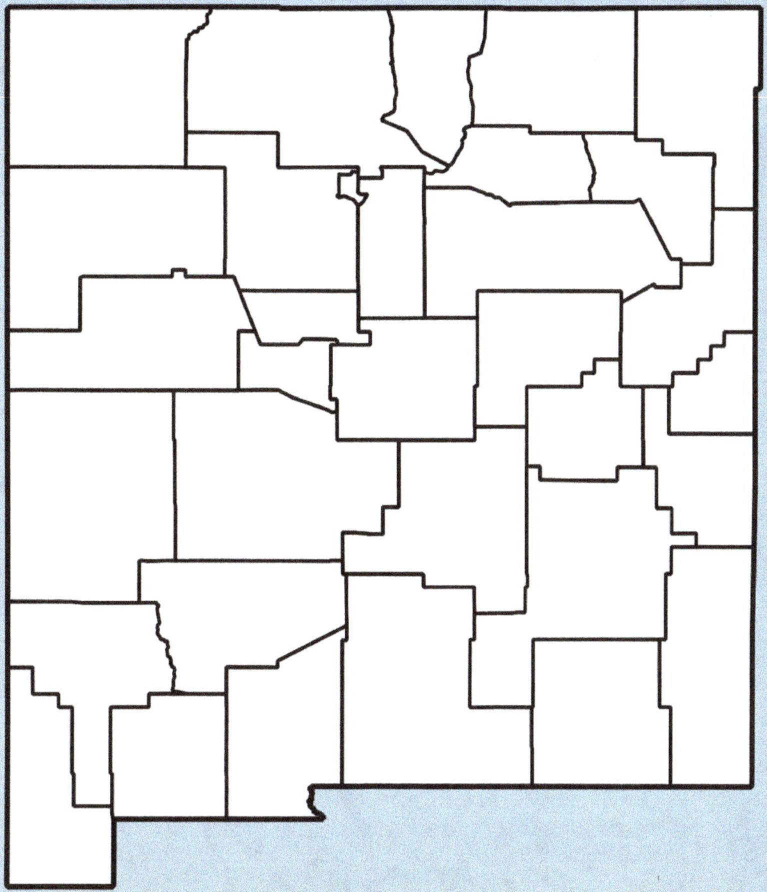

NEW YORK

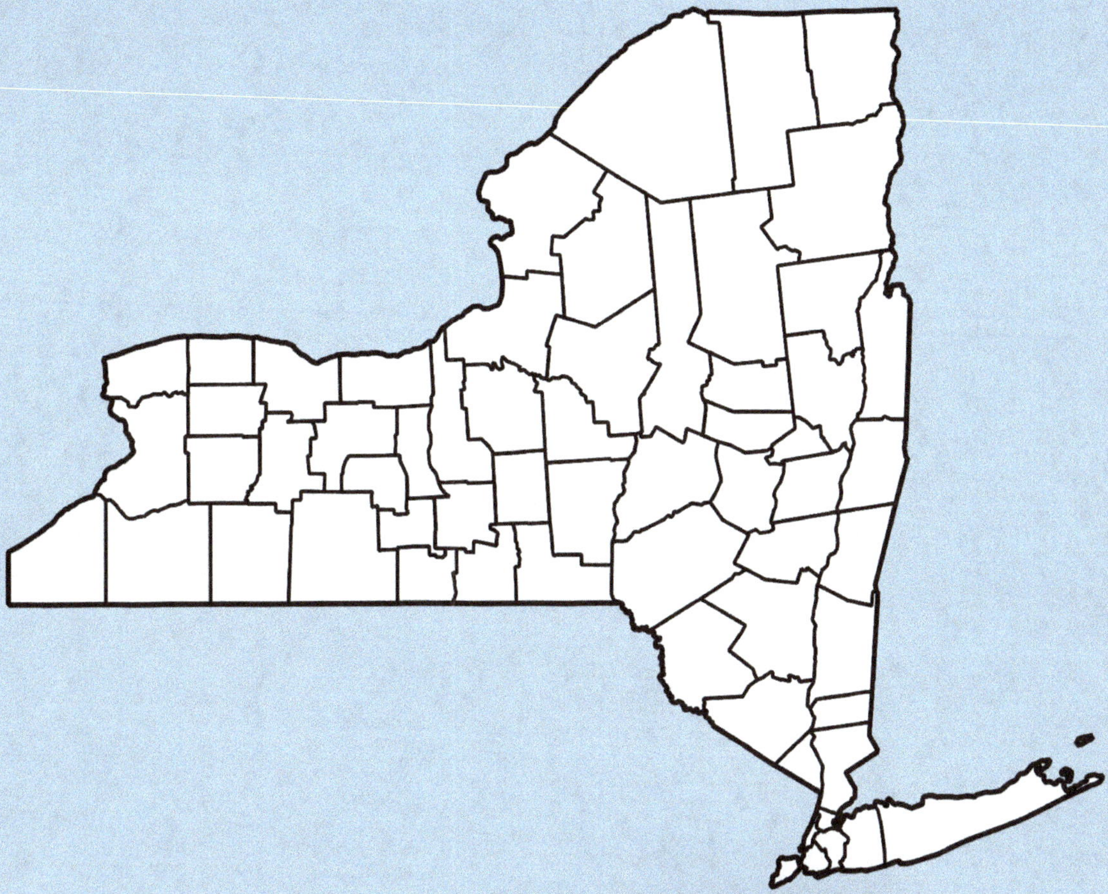

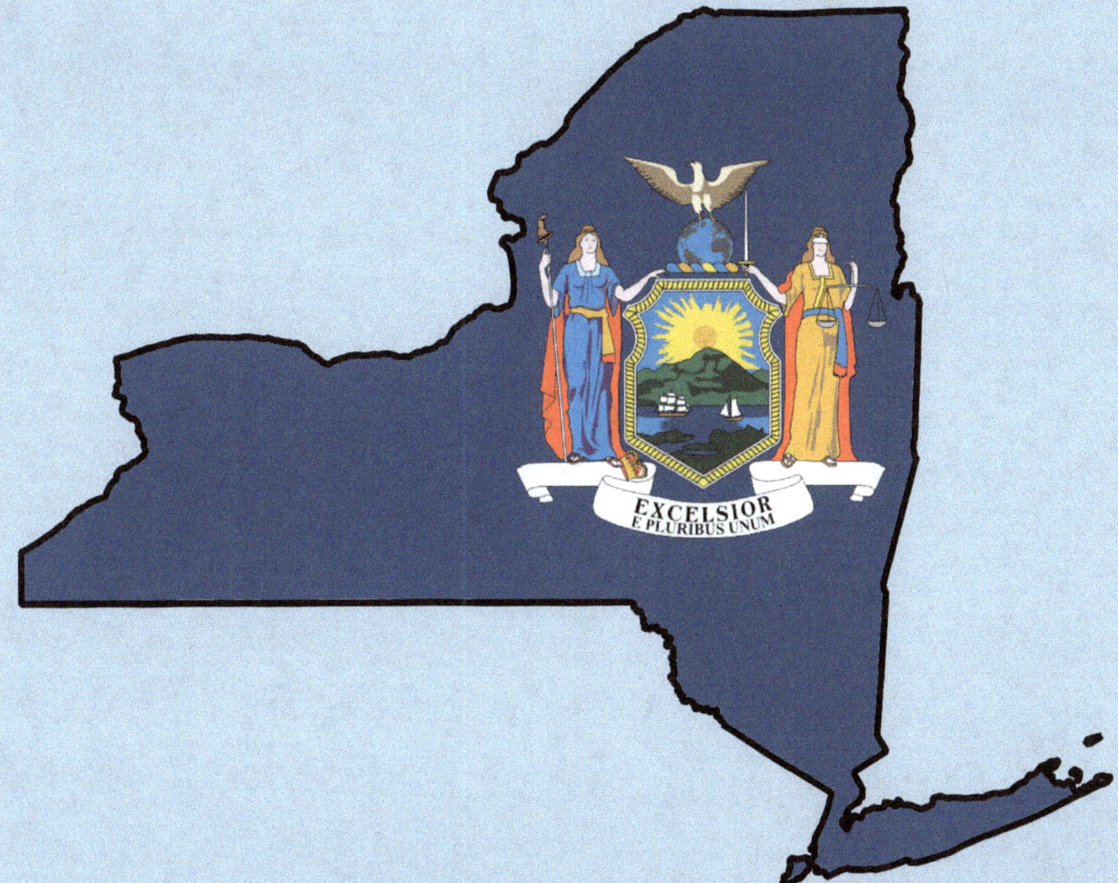

NORTH CAROLINA

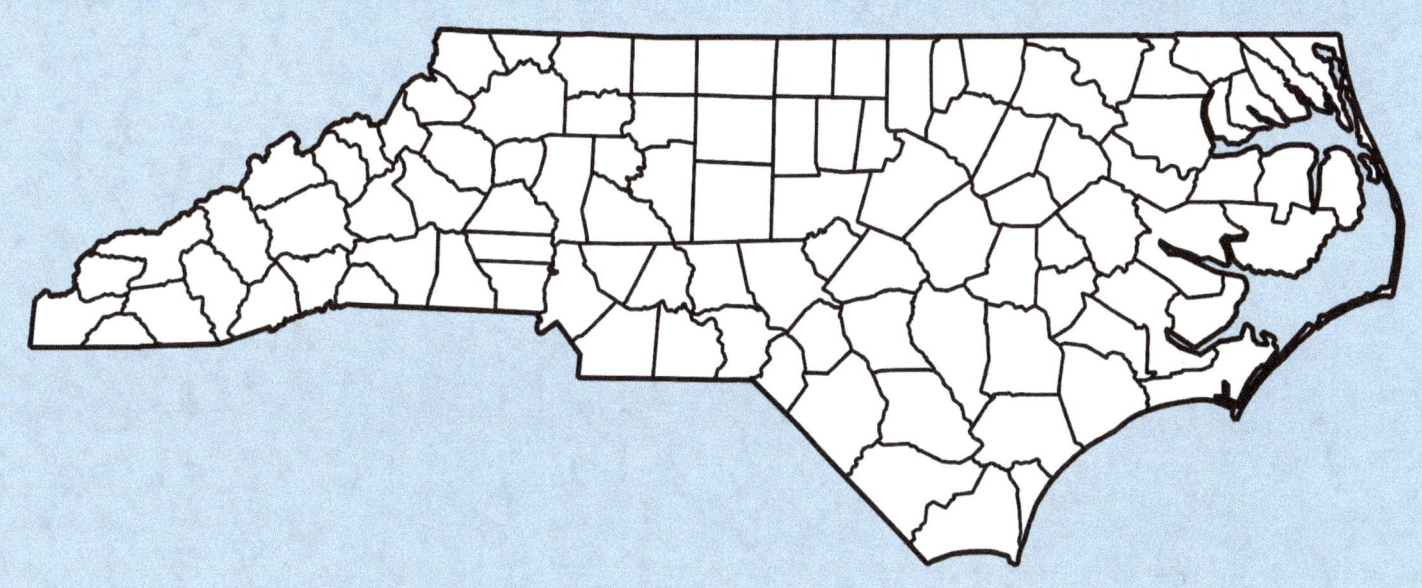

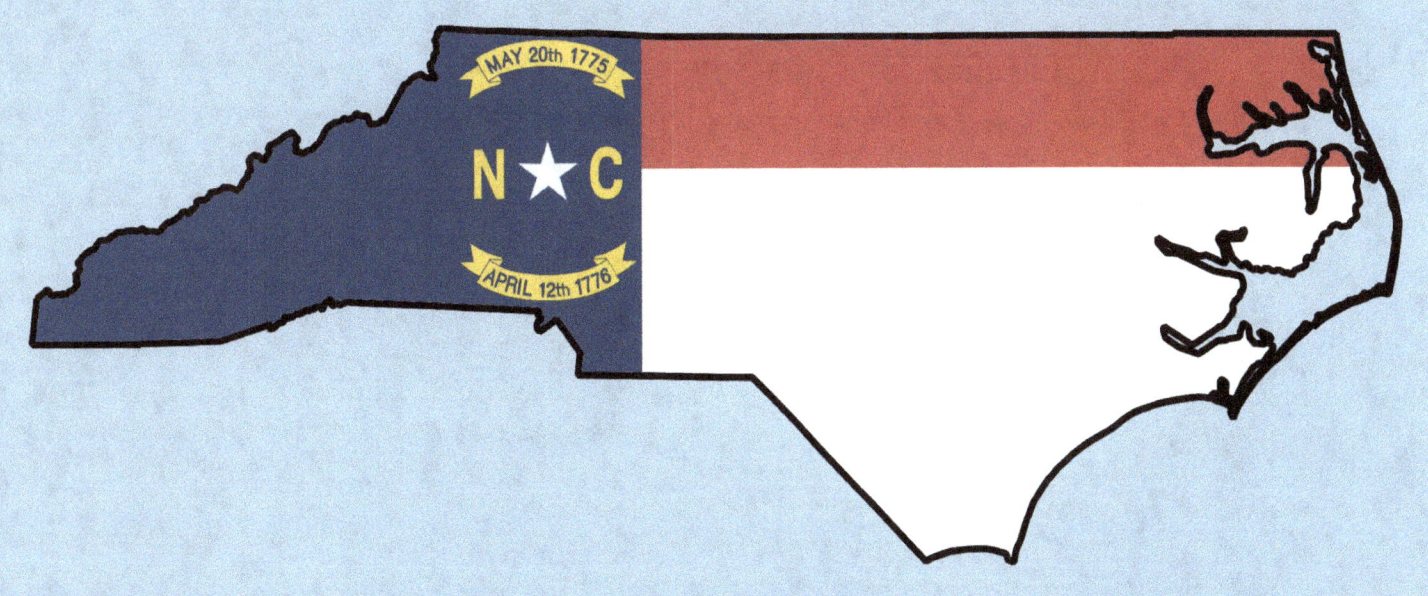

NORTH DAKOTA

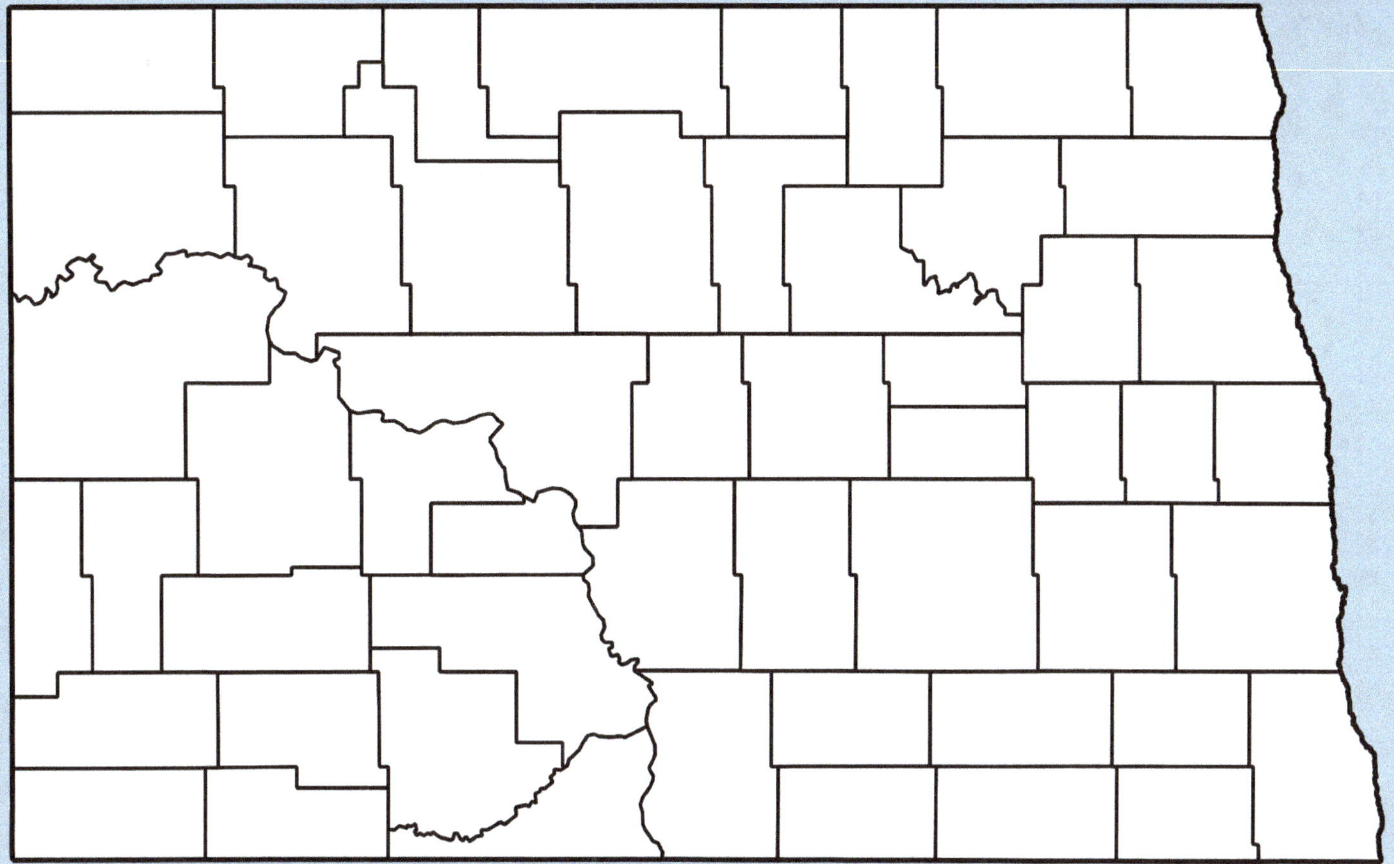

OHIO

OKLAHOMA

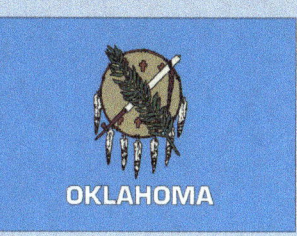

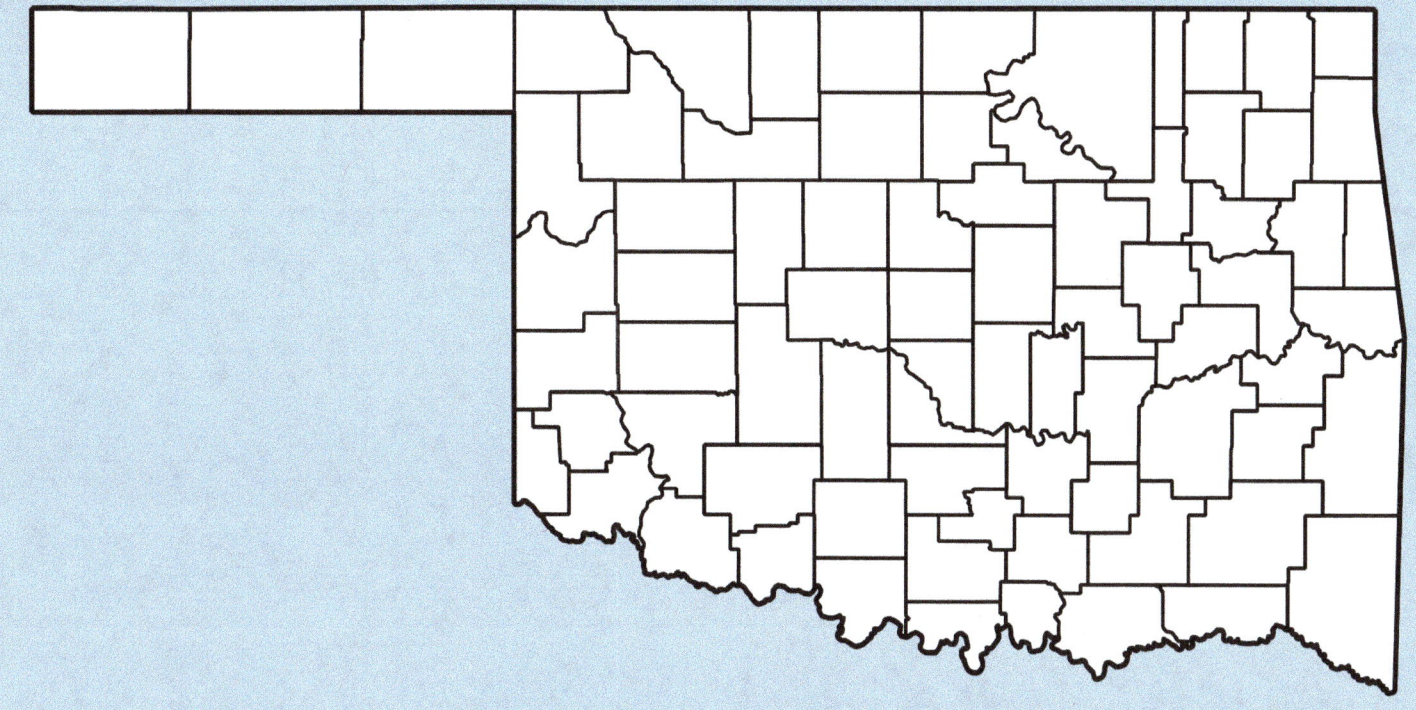

OREGON

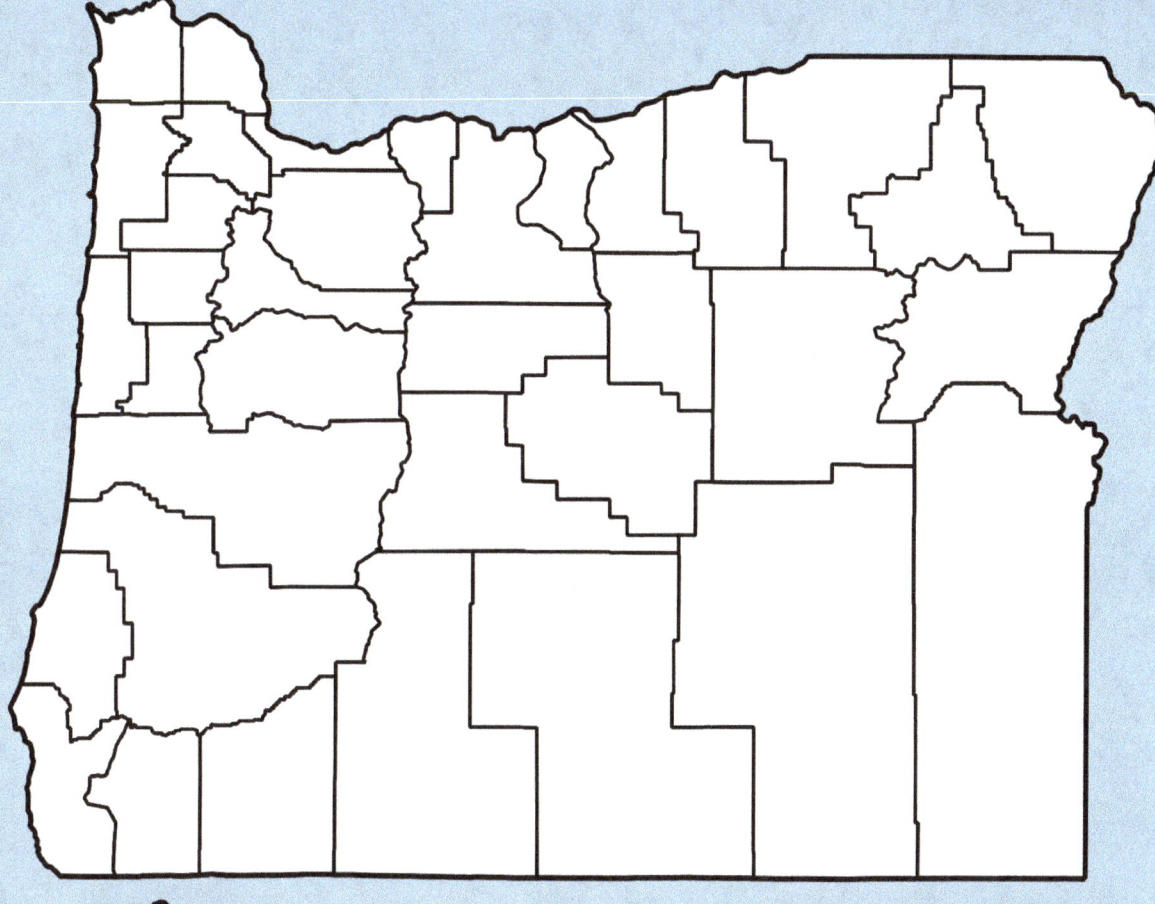

PENNSYLVANIA

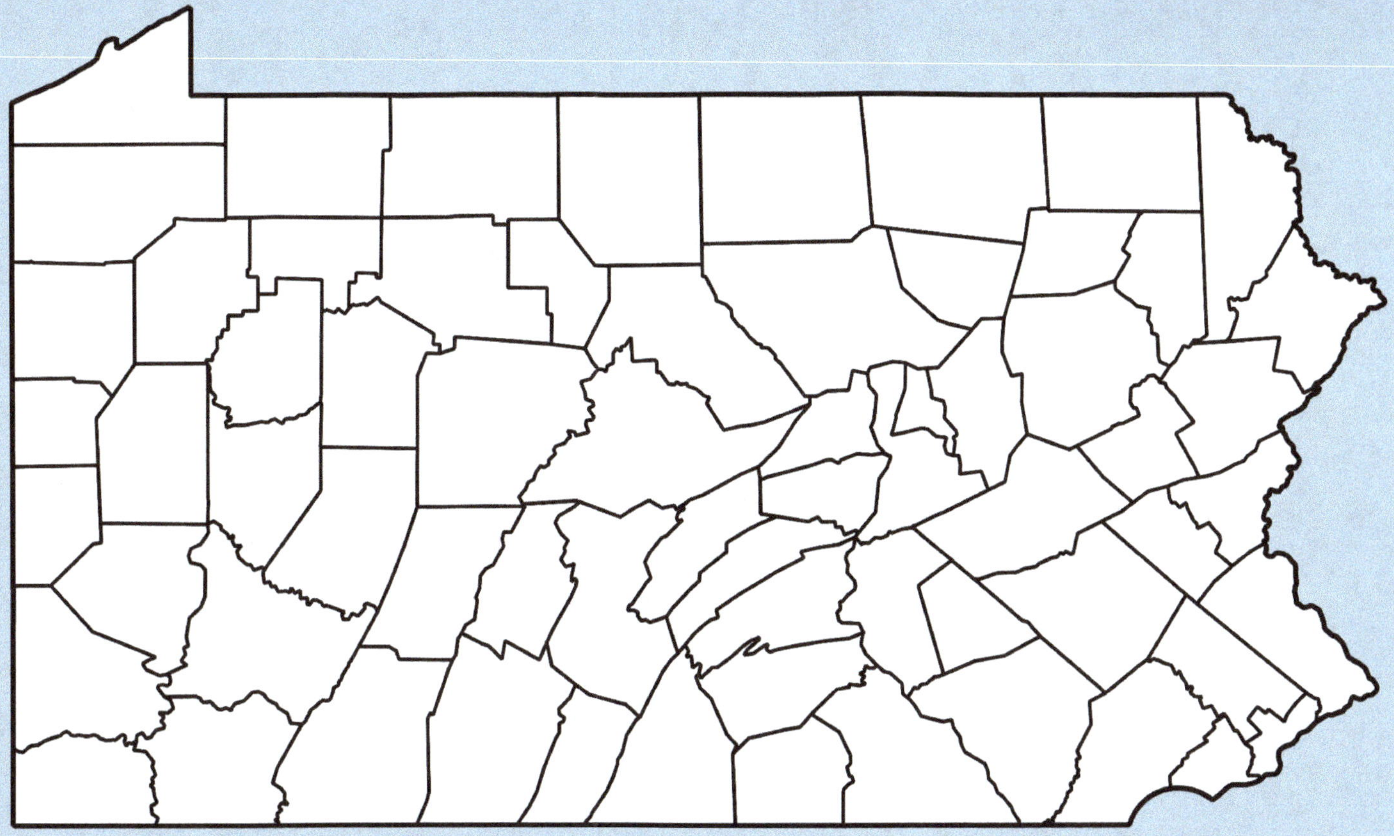

RHODE ISLAND

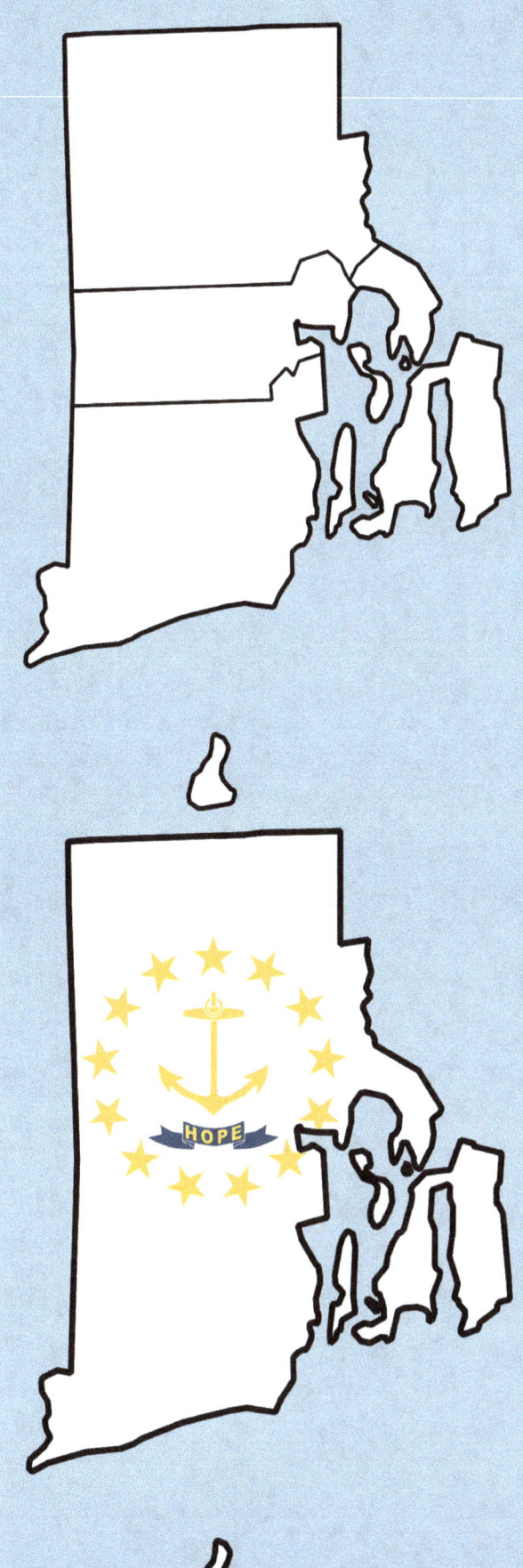

SOUTH CAROLINA

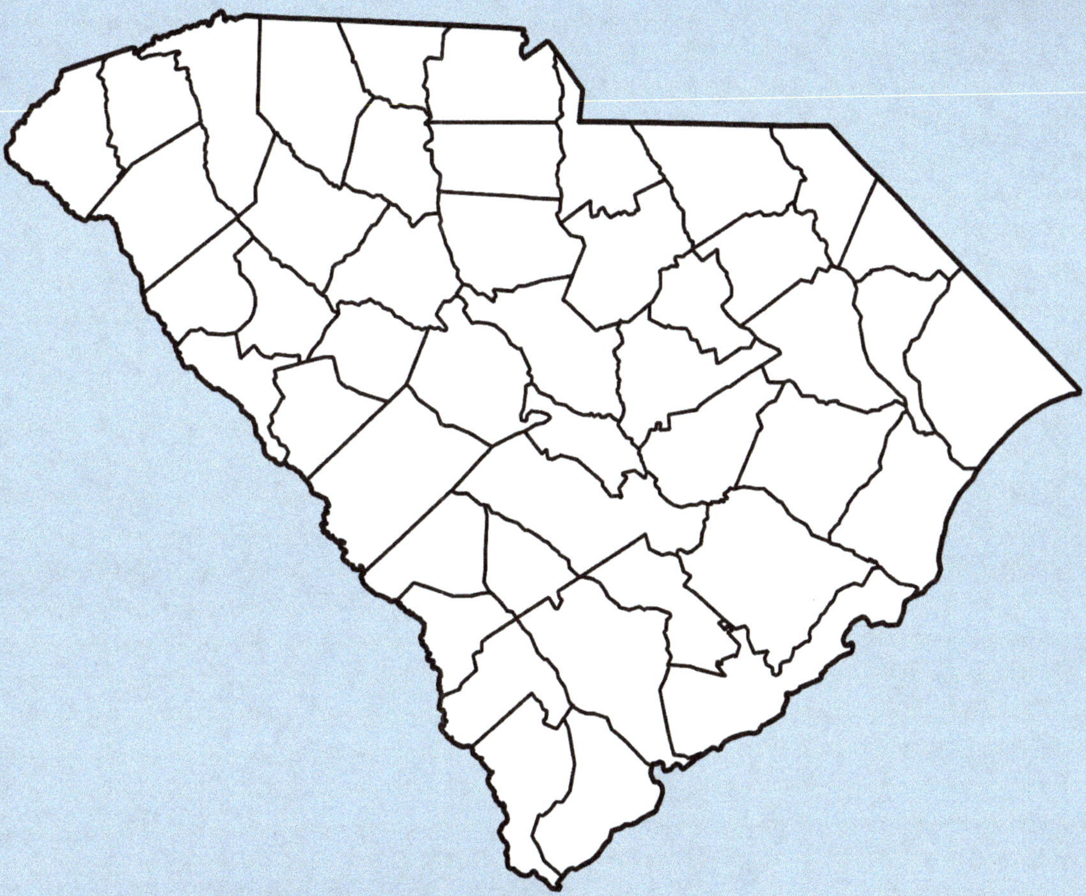

SOUTH DAKOTA

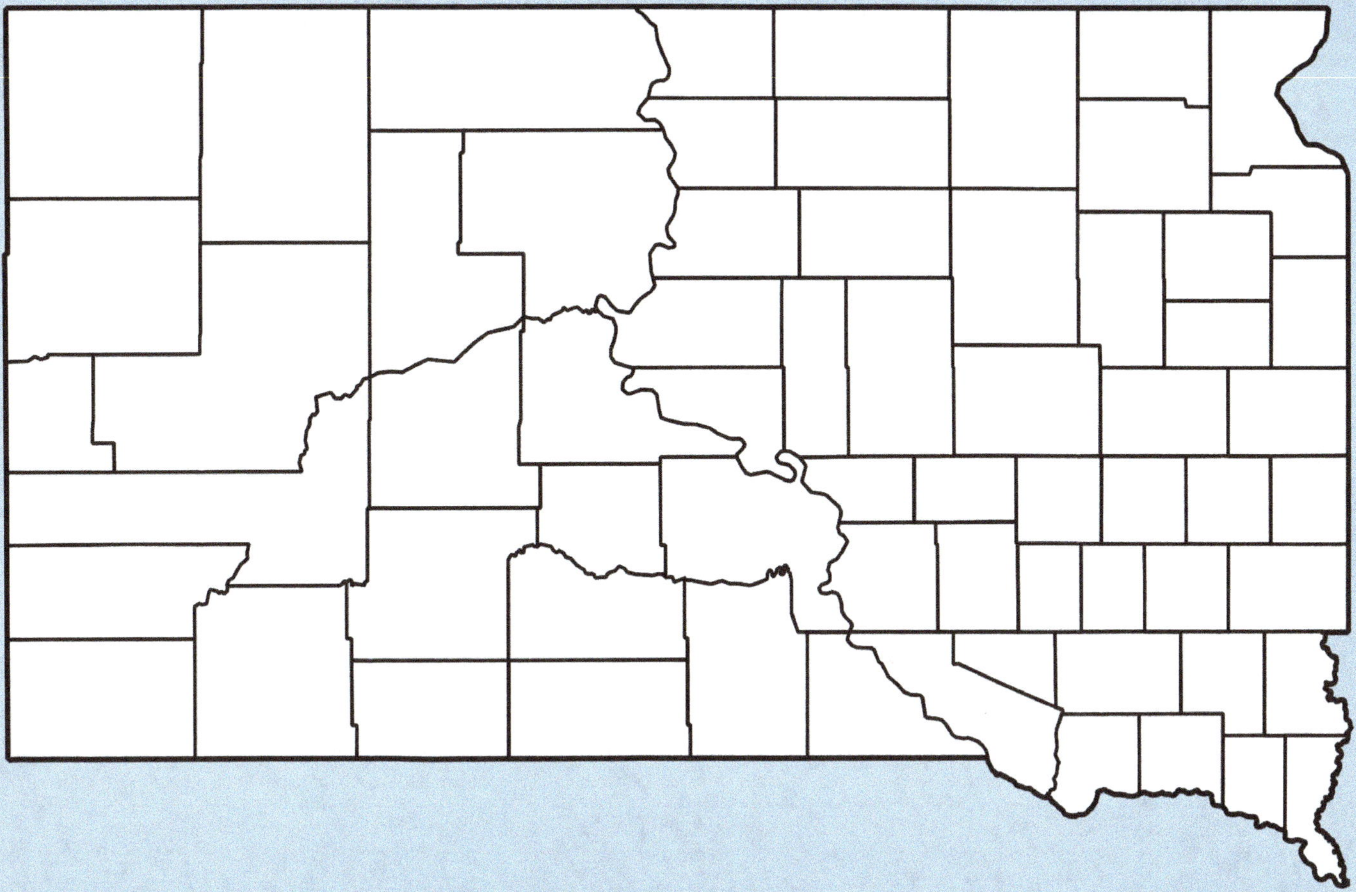

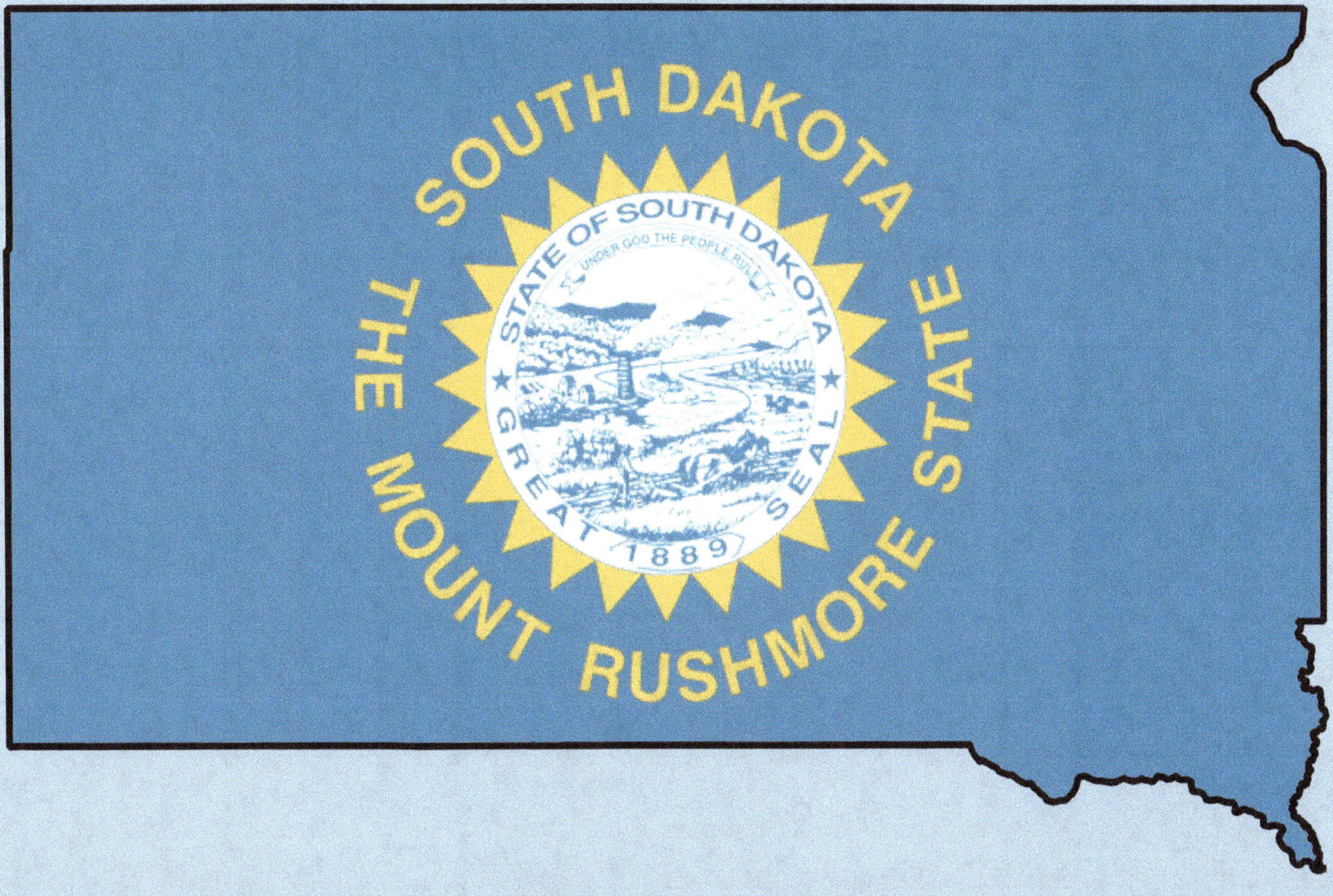

TENNESSEE

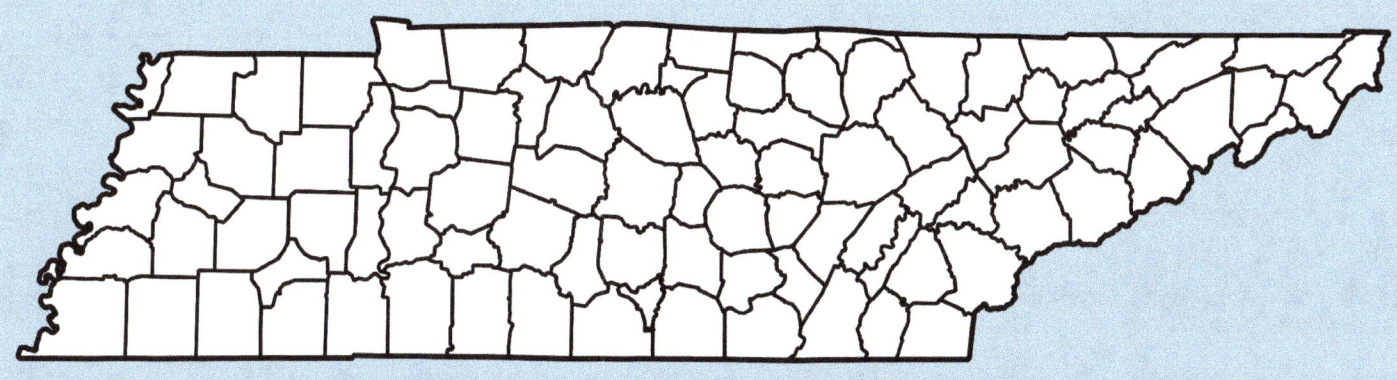

TEXAS

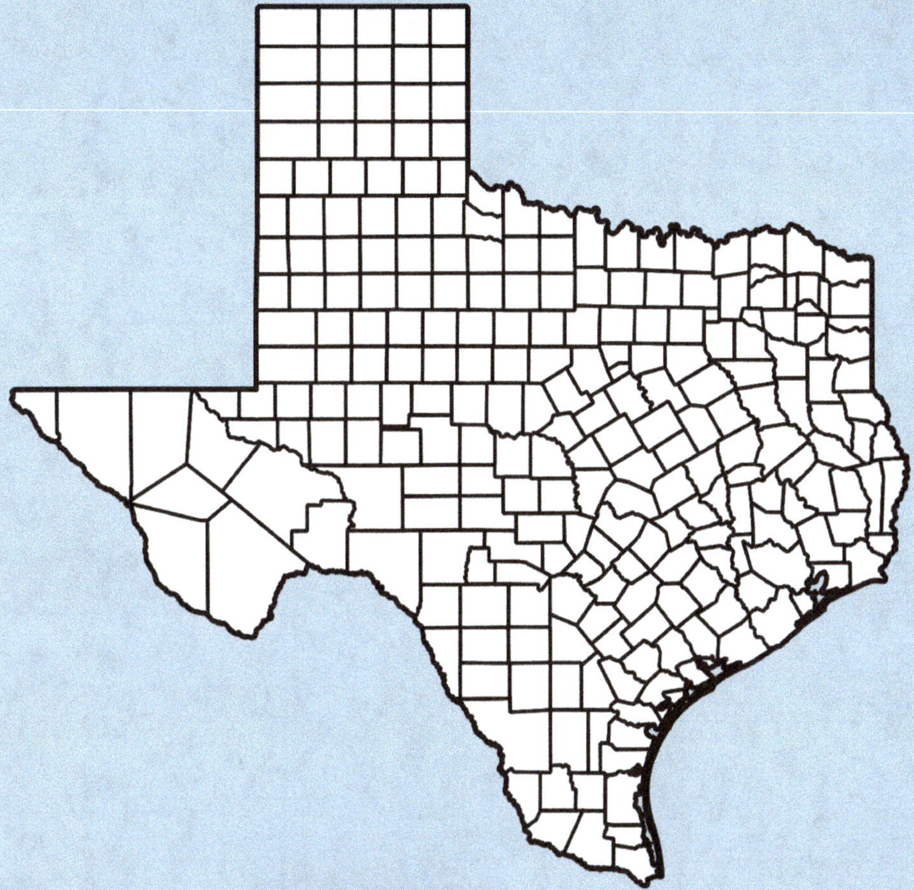

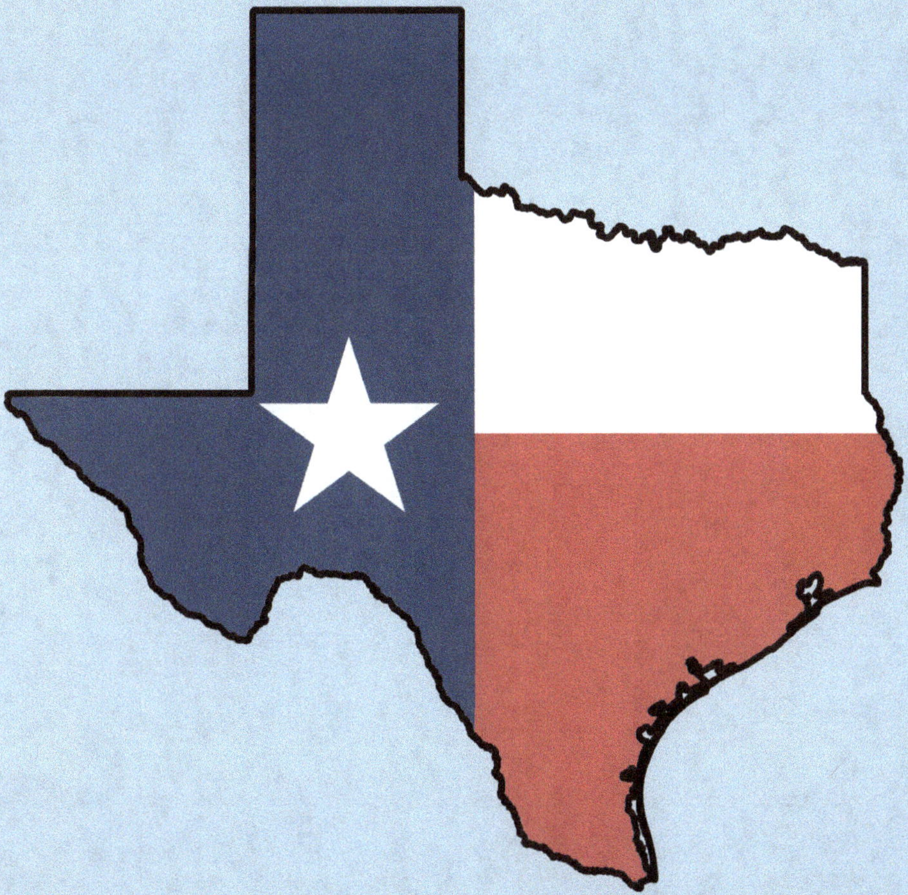

UTAH

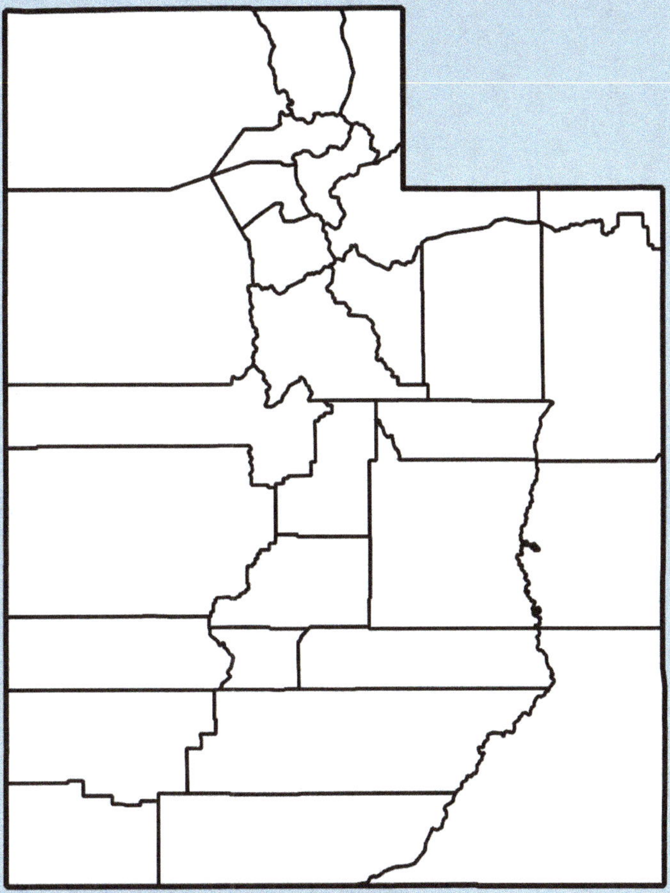

VERMONT

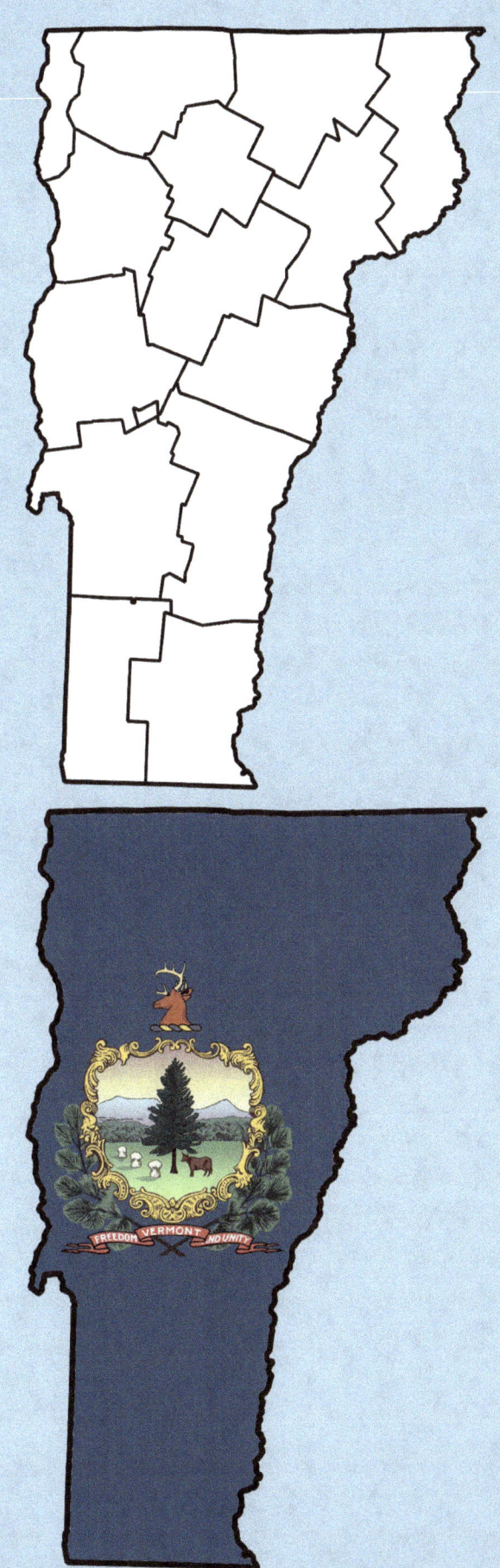

VIRGINIA

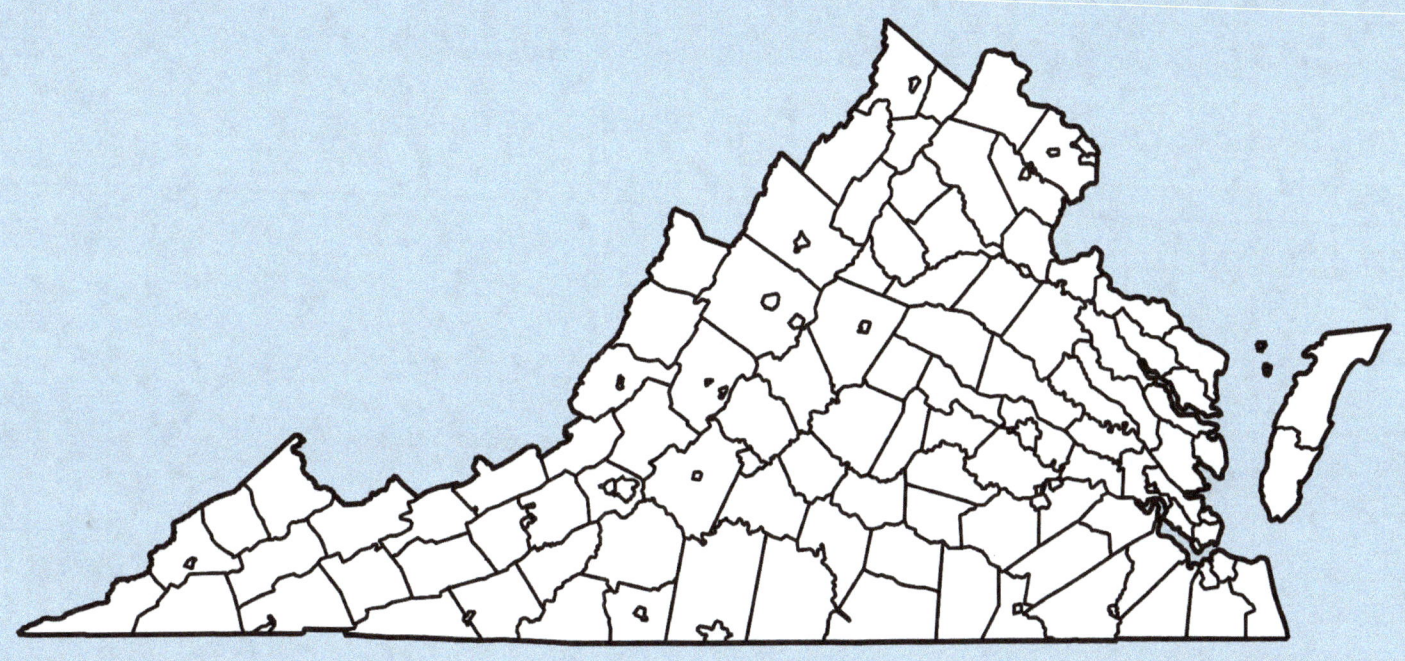

WASHINGTON

WEST VIRGINIA

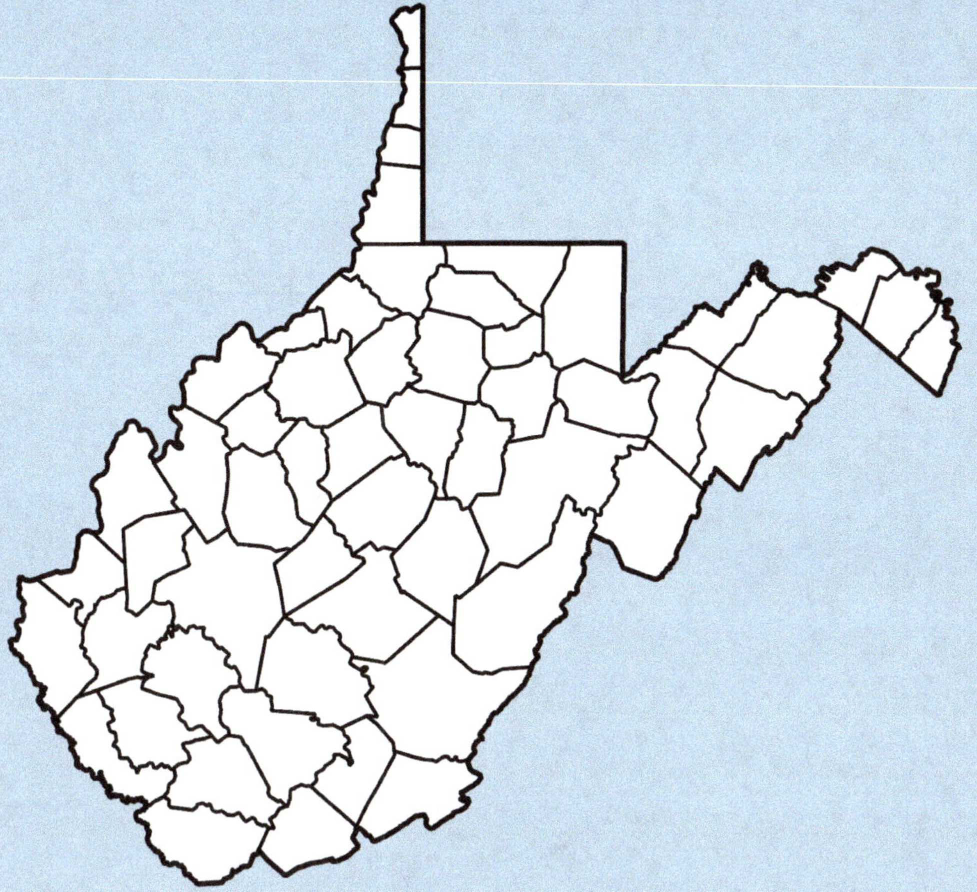

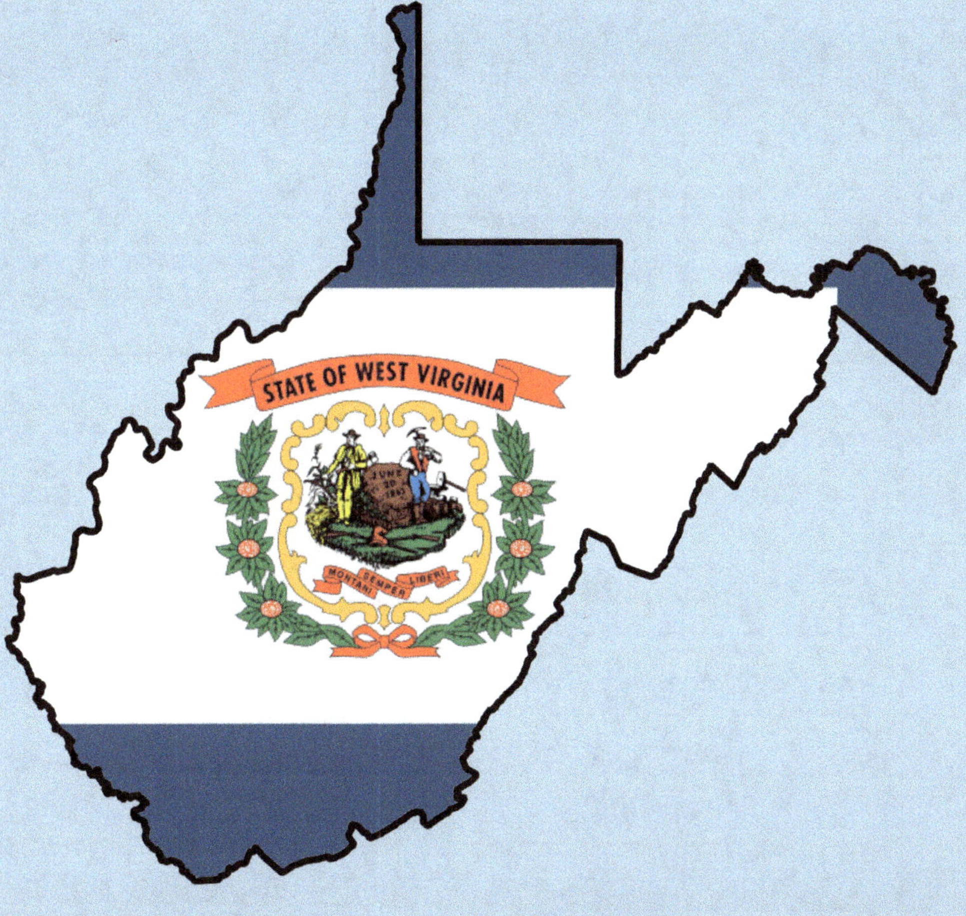

WISCONSIN

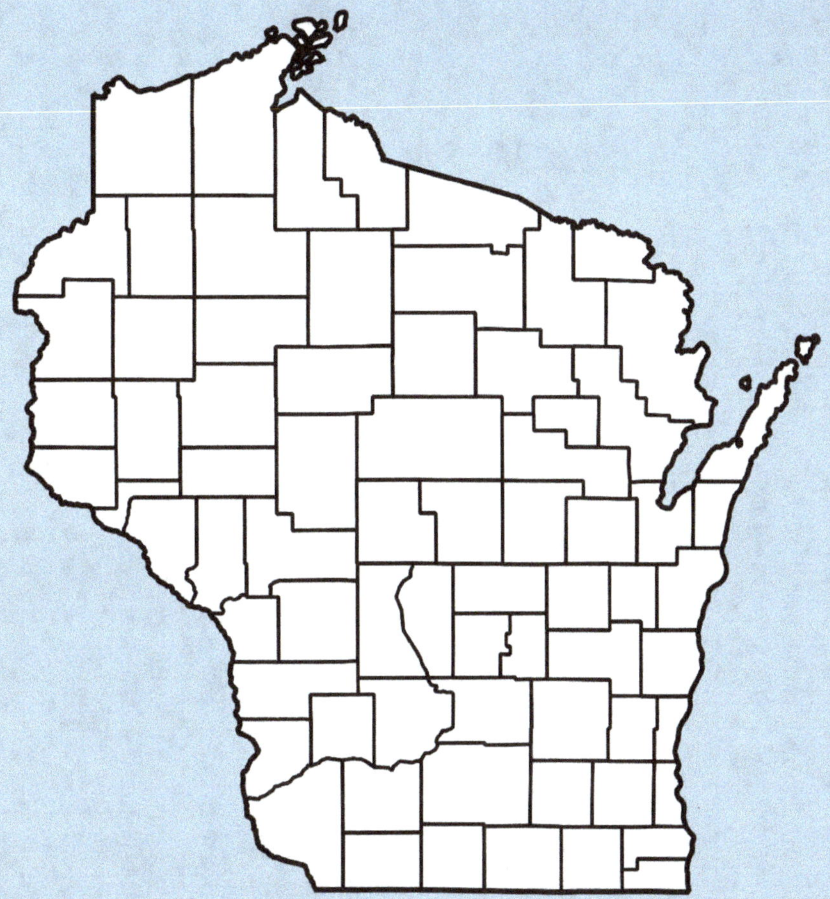

WYOMING

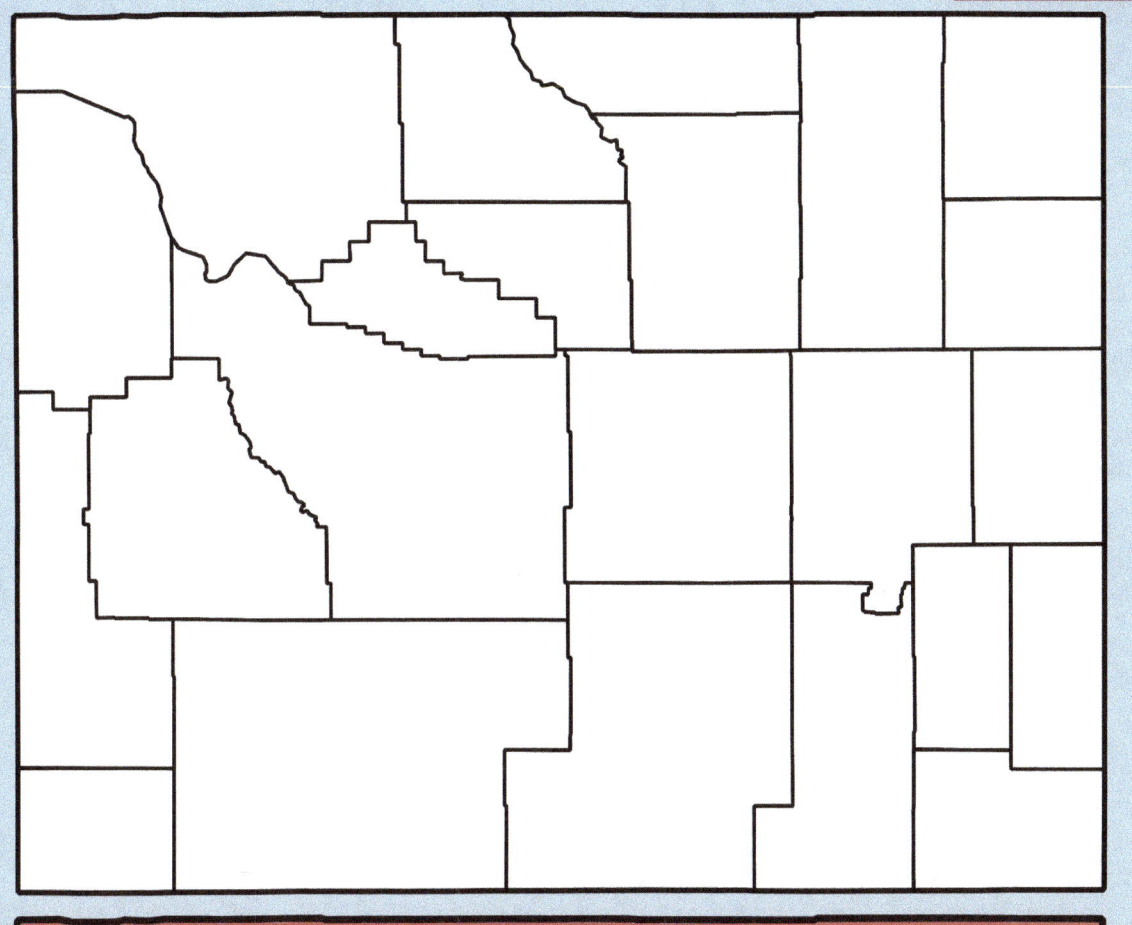